EVESHAM & MALVERN HILLS COLLEGE

Badisches Landesmuseum Karlsruhe, Museum beim Markt
21. Dezember 2000 bis 18. März 2001
Deutsches Textilmuseum Krefeld
8. April 2001 bis 1. Juni 2001
Landesmuseum für Vorgeschichte Dresden
9. August bis 11. November 2001

Katharina Thomas (ed.)

FELT

Art,
Crafts and
Design

ARNOLDSCHE ART PUBLISHERS

Katharina Thomas (Hrsg.)

FILZ

Kunst,
Kunsthandwerk
und Design

ARNOLDSCHE ART PUBLISHERS

Jury
Selection committee

Liesbeth Crommelin, Amsterdam
Rosmarie Lippuner, Lausanne
Jan Walgrave, Antwerpen
Babette Küster, Leipzig
Peter Schmitt, Karlsruhe

Leihgeber
Außer den beteiligten Künstlern haben folgende
Personen und Institutionen Leihgaben für die
Ausstellung zur Verfügung gestellt:

Lender
Besides the artists invited the following
persons and institutions lent to the exhibition:

Mary E. Burkett, Cockermouth, Cumbria
Museum Boijmans van Beuningen, Rotterdam
Cappellini S. p. A., Arosio (Como)
Sammlung Froehlich, Stuttgart
Stiftung Schloß Moyland – Joseph Beuys Archiv
des Landes Nordrhein-Westfalen
People & Nature e.V., Berlin
Edition Staeck, Heidelberg

Weitere Leihgeber wollen ungenannt bleiben

Inhalt Contents

Vorwort

Das Deutsche Textilmuseum Krefeld, das Badische Landesmuseum Karlsruhe, das Landesmuseum für Vorgeschichte in Dresden – drei Museen mit je eigener Prägung haben bei einem Thema zusammengefunden, das sie alle, den eigenen Sammlungsbestand bereichernd oder kontrastierend, auf jeweils unterschiedliche Weise berührt.

In Krefeld etwa treffen zwei Materialien textiler Produktion aufeinander, wie sie unterschiedlicher kaum sein könnten. Die Stadt war über mehr als zwei Jahrhunderte berühmt für ihre Seidenweberei, für die nicht nur das kostbarste aller textilen Materialien kennzeichnend ist, sondern ebenso hochkomplizierte technische Gerätschaften, die sich von den Anfängen des ersten Zugwebstuhls in sehr kleinen Schritten über große Zeiträume hinweg bis zur Jacquardmaschine des 19. Jahrhunderts entwickelten. Immer waren es die vornehmsten Personen, für die Seidengewebe angefertigt wurden – für den täglichen Gebrauch fanden solche Stoffe keine Verwendung. Ganz anders der Filz, dessen Herstellung sicher zu den ältesten Textiltechniken überhaupt gehört, auch wenn man darüber mangels ausreichender Nachrichten und wegen der raren erhaltenen Beispiele wenig Genaues sagen kann. Eine wichtige Rolle spielte er seit Urzeiten bei den Nomaden Zentralasiens. Denn um ihn herzustellen, braucht man wenige Hilfsmittel und ist an keinen Ort gebunden. Dazu kommt seine wichtigste Eigenschaft: er bietet Schutz nicht nur gegen Kälte, sondern gleichermaßen gegen übergroße Wärme. Die Zelte der Mongolen sind hierfür das beste Beispiel, denn die Filzbahnen dieser runden Jurten halten im Winter wie im Sommer ein gleichmäßig angenehmes Klima. Auch Kleidung – Mäntel, Hüte, Socken und Schuhe – stellte man aus Filz her; ebenso Decken, die als Bodenbeläge und Wandbehänge, aber auch zur Bestattung der Toten verwendet wurden.

Während es keinerlei Beweise oder Nachrichten darüber gibt, daß man in Amerika, Afrika oder Australien gefilzt hätte, sind für den Mittelmeerraum Belege aus dem 6. Jahrtausend v. Chr. aus Çatal Hüyük, einem Gebiet in der heutigen Türkei, erhalten. Auch die Griechen und nach ihnen die Römer haben sich die schützenden Eigenschaften des Filzes für Kleidung und Militärausrüstung zunutze gemacht. Die kostbarsten frühgeschichtlichen Zeugnisse aber kommen aus den zwischen 600 und 200 v. Chr. angelegten Gräbern von Pazyryk im Altai-Gebirge, wo sie in 1600 m Höhe im Permafrost überdauerten. Sie werden heute in der Eremitage von St. Petersburg aufbewahrt. Die Lebendigkeit und Vielseitigkeit der Muster dort gefundener Filze, ihre Kraft und Eleganz sowie die strahlende Leuchtkraft ihrer Farben sind ganz und gar verblüffend. Vor allem ein Behang mit zwei Reihen von Reitern, die einer Gottheit huldigen, ist in seiner kühnen Farbigkeit und kraft der dynamischen Darstellung von Tier und Reiter auch für den modernen Betrachter von atemberaubender Schönheit.

Solche Werke heben das Filzen, das zunächst als einfache Technik vorgestellt wurde, weit hinaus über jede Form primitiver Handwerkskunst. Damit steht es der Seidenweberei in nichts nach. Wie diese ist es eine faszinierende Kunst, die über Jahrhunderte und Jahrtausende schöpferische Hochleistungen hervorgebracht hat. An sie knüpfen heutige Künstler an. Es sind vor allem Frauen, die das Filzen, das unmittelbare Arbeiten mit der natürlichen oder gefärbten Faser, als künstlerisches Ausdrucksmittel entdeckt haben und in dieser uralten Technik gegenwärtige Erfahrungen verarbeiten. Auf der anderen Seite gibt es seit Joseph Beuys Künstlerinnen und Künstler, die weniger am Prozeß der Herstellung, sondern mehr an den unterschiedlichen emotionalen Eigenschaften des Filzes interessiert sind. Beides zusammen ergibt ein Gesamtbild zeitgenössischer Filzkunst, das sich in jedem der drei Museen vor einer je eigenen Folie präsentiert.

Prof. Dr. Harald Siebenmorgen
Badisches Landesmuseum Karlsruhe

Dr. Brigitte Tietzel
Deutsches Textilmuseum Krefeld

Dr. Judith Oexle
Landesmuseum für Vorgeschichte Dresden

Foreword

The Deutsches Textilmuseum Krefeld, the Badisches Landesmuseum Karlsruhe, the Landesmuseum für Vorgeschichte in Dresden – each one of them a museum with a distinctive character of its own – have collaborated on a joint project which relates to each museum in different ways since it supplements or contrasts with their own collections.

For Krefeld it entails the meeting of two materials of textile production as different as they can be. For more than two centuries the city was famous for silk weaving. Silk is not only the most costly of all textiles; weaving it also requires highly sophisticated equipment, evolving from the first primitive handlooms very gradually over a large period of time to the invention of the Jacquard machine in the 19th century. Silk fabric was invariably made for the rich and elegant; such materials were not used for daily wear.

Felt, however, is a different matter altogether. Felting must surely be one of the earliest techniques for making textiles even though little can be said about its origins with any accuracy since written records are scanty and few examples of really ancient felt are extant. Felt has played an important role in the nomadic cultures of Central Asia since time immemorial because few aids are needed for making it and felting is not bound to a place. Moreover, its most important quality comes into play in this connection: felt not only affords protection from cold but also insulates against extreme heat. Mongol tents are the best example of this since the rolls of felt from which these round yurts are made maintain an even, pleasant temperature in them year round, in summer and in winter. Clothing – coats, hats, socks, and shoes – has also always been made of felt; so have blankets, which are used to carpet yurts and cover their walls, but also to bury the deceased.

Although there is no archaeological evidence nor any oral tradition to support the assumption that felt was made in prehistoric America, Africa or Australia, there is, on the other hand, evidence for felting in the Mediterranean, dating from the 6th millennium BC in the region of Çatal Hüyük, in what is now Turkey. The ancient Greeks, followed by the Romans, also took advantage of the insulating and protective properties of felt for making clothing (particularly for military uses).

The most valuable early historical evidence for felting, however, dates from between 600 and 200 BC, in the graves at Pazyryk in the Altai Mountains, where felt was preserved in permafrost at an elevation of 1600 m. These finds are now in the Hermitage in St Petersburg. The liveliness and variety of patterns on felt materials found there, their vibrancy and elegance as well as the brilliance of their colours are utterly astonishing. A prime example is a wall hanging with two friezes of horsemen who are worshipping a deity. Its bold colour and the forcefulness of the dynamic representations of both animals and riders on it are for the modern viewer breathtakingly beautiful.

Such works of art elevate felting, which was at first decried as merely a simple technique, high above any form of primitive craft. It is in every way the equal of silk weaving. Like silk weaving, felting is a fascinating art which for centuries and even millennia has brought forth superlative creative achievements. Today's artists working in felt are part of that tradition. Most of those who have discovered felting as their medium of creative expression, working directly with the natural or dyed fibres and dealing with present experience in this ancient technique, are women. Their work is paralleled by the line started by Joseph Beuys, where the focus of interest is not on the process of making felt but on the various properties of felt itself. Both aspects of felting taken together amount to a sweeping overview of contemporary art in felt, which is presented in each of the three museums against its own proper background.

Prof. Dr. Harald Siebenmorgen
Badisches Landesmuseum Karlsruhe

Dr. Brigitte Tietzel
Deutsches Textilmuseum Krefeld

Dr. Judith Oexle
Landesmuseum für Vorgeschichte Dresden

Filz – Kunst, Kunsthand-werk, Design

Peter Schmitt

Filz ist »in«. Man könnte versucht sein zu fragen, ob es mit der zunehmenden sozialen Kälte in unserer Gesellschaft zu tun hat, daß ein Werkstoff wiederentdeckt wird, der nicht erst seit Joseph Beuys Sinnbild für Wärme, Leben und Geborgenheit ist. Jedenfalls findet Filz seit den späten 80er Jahren bei Künstlern, vor allem bei Künstlerinnen, weltweit verstärktes Interesse, wobei es nun weniger das industriell gefertigte Produkt ist, mit dem schon in den 60er Jahren die beiden Künstler Joseph Beuys und Robert Morris auf ganz unterschiedliche Weise gearbeitet haben, sondern häufig der Prozeß des Filzens selbst, der für die künstlerische Aussage bestimmend ist.

Für die jüngere Generation von Künstlern, die sich mit Filz beschäftigen, kann Joseph Beuys noch immer als eine Art Vaterfigur angesehen werden. Er hat diesem Stoff ebenso wie anderen Materialien, die in seinem Werk eine Rolle spielen, symbolische Bedeutung verliehen und auf seine Herkunft aus der Kultur der Nomaden verwiesen, die in seinem Denken einen zentralen Platz einnahm. Auch wenn Beuys, seinem Verständnis von Pädagogik entsprechend, keine direkte Nachfolge finden konnte, ist doch nicht zu übersehen, daß er der Entdeckung des Filzes als künstlerischem Medium den Boden bereitet hat. Deutlich auf ihn Bezug nimmt Ines Tartler mit ihrem Filzanzug »alles wollen nach innen schlagen« und ihrer Filzschürze mit Unterhose, einem metaphorischen Kleidungsstück, das Handhabung provoziert und als autonomes Objekt zugleich in Frage stellt.

Der Einfluß von Robert Morris erstreckt sich weniger auf die Verwendung von Filz, obwohl die »felt pieces« einen wichtigen Komplex seines Werks darstellen, als vielmehr auf das Konzept der Anti-Form, das er nicht zuletzt mit diesem Material demonstriert hat. Ihn interessierte, welche Formen dieser schwere, flexible Stoff unter Einwirkung der Schwerkraft und anderer Faktoren ausbildete, wenn er Einschnitte in die Filzbahn vornahm und diese dann mit Haken an der Wand befestigte. Dabei gab der Künstler nur einige Variablen vor, verzichtete aber darauf, den Prozeß der Formgebung zu steuern und vollzog damit einen weitreichenden Paradigmenwechsel, indem er sich aus seiner Rolle als Maler oder Bildhauer verabschiedete.

Die ursprüngliche Materialität des Filzes, mit der Morris noch arbeitete, ist weitgehend aufgehoben in den Arbeiten von Ernst Köster, der ihm durch Verkleben mehrerer Platten seine Flexibilität nimmt und ihn in ein starres Material verwandelt, aus dem er seine technoiden Körper aussägt. Auch bei Ulrike Hein ist die Verwendung von Filz nicht auf den ersten Blick zu erkennen. Sie bildet moderne Kriegswaffen täuschend in Filz nach, übersetzt das Starre, Tödliche in Weiches, Warmes, ohne daß die Objekte durch die Abformung in einem fremden, unbrauchbaren Werkstoff ihren Schrecken einbüßten.

Spezifische Materialqualitäten geben auch den formal und farbig aufs äußerste reduzierten Wand- und Bodenarbeiten von Maisa Tikkanen und Verena Gloor ihren besonderen Charakter. Beide Künstlerinnen verarbeiten Einflüsse der Minimal Art, deren abweisende Strenge durch das warme Material und das Sichtbarmachen des handwerklichen Prozesses jedoch unterlaufen wird. Ihre Objekte besetzen den Raum nicht nur, sondern geben ihm eine meditative Mitte. Eine ähnliche Qualität weisen die Arbeiten von Gunilla Paetau Sjöberg mit ihren flirrenden Schwarztönen auf. Ingeborg Verres dagegen verbindet den amorphen, weichen Filz in ihrem »Lichtfeld« kontrastierend mit starren, von innen leuchtenden Glasröhren zu einer künstlichen Landschaft von stiller Unheimlichkeit.

Mehr als an seinen plastischen Möglichkeiten scheinen zeitgenössische Künstlerinnen am Filz als Bildträger interessiert zu sein. Ganz wörtlich wird diese Funktion von Ute Lindner genommen, die für ihre Serie »Belichtungszeiten« Wandbespannungen aus rotem Filz verwendet, die in der Kassler Gemäldegalerie einmal als Untergrund für die dort ausgestellten Bilder gedient haben. Lange dem Licht ausgesetzt, haben diese Filzbahnen ihre ursprüngliche Farbe fast ganz verloren. Auf dem ausgeblichenen Grund zeichnen sich nur noch die Umrisse der einmal dort aufgehängten Bilder als dunklere Flächen ab, Schatten von Bildern, die nun selbst zu Bildern werden.

Bildträger in einem direkten Sinn ist der Filz bei Silja Puranen, die in sehr persönlichen Formulierungen Rolle und Schicksal der Frau umkreist, indem sie Fotografien aus dem eigenen Leben auf Filz überträgt und sie mit anderen »weiblichen« Materialien wie Spitzen u.ä. verbindet. Denise Bettelyoun nennt ihre Arbeiten » Filztafelbilder«; ihre Tierdarstellungen, die sie in die naturweiße Unterlage einfilzt, verweisen wohl nicht zufällig auf Zeichnungen von Joseph Beuys, auf den nicht zuletzt auch durch das Material Filz Bezug genommen wird.

Einige Künstlerinnen entwickeln ihre Bilder direkt aus dem Prozeß des Filzens heraus. Die farbigen Flächen und Motive sind also nicht zusammengenäht, sondern miteinander verfilzt. Beispiele sind die trotz ihrer sehr freien Muster noch als Gebrauchstextilien, Teppiche oder Wandbehänge, erkennbaren Arbeiten von Inge Evers, May Jacobsen Hvistendahl und Turid Kvalvåg sowie die doppelseitigen Filzbilder von Claudia Merx.

Voraussetzung für diese Arbeiten ist die Wiederentdeckung der handwerklichen Filztechniken, wie sie besonders bei den Nomadenvölkern Zentralasiens lebendig geblieben sind. Sie ist vor allem den ethnologischen Arbeiten von Mary E. Burkett zu verdanken. Fast ebenso wichtig sind die Bemühungen von Mari Nagy und István Vidák, die von der ungarischen Volkskunst ausgehend ihre Untersuchungen bis in die Türkei und nach Zentralasien ausdehnten. Sie beschränkten sich nicht auf das Sammeln und Dokumentieren, sondern haben als versierte Handwerker die Techniken des Filzens selbst praktisch erprobt. Obwohl sie an traditionellen, in der Volkskunst wurzelnden Motiven festhielten, haben sie als Lehrer eine Wirkung entfaltet, die weit über die Grenzen ihres eigenen Landes hinausreicht. Unter dem von Inge Evers übernommenen Motto »filzt die Welt zusammen« bildete sich ausgehend von den Workshops in Kecskemét eine Gemeinde, für die der Prozeß des Filzens von elementarer Bedeutung ist, weil er, wie die Äußerungen zahlreicher Beteiligter belegen, künstlerisch-handwerkliche und meditative Aspekte in sich vereinigt.

Es sind fast ausschließlich Frauen, die auf diesem Feld tätig sind, das eine vom offiziellen Kunstbetrieb weitgehend ausgesparte Enklave darstellt. Dies hat der Filz mit anderen künstlerischen Ausdrucksformen gemein, die an bestimmte, meist aus dem Kunsthandwerk kommende Materialien gebunden sind. Dem Filz scheint darüber hinaus eine Qualität innezuwohnen, die viele Künstlerinnen in ihm ein Medium erkennen lässt, ihre Situation angemessen zu reflektieren.

Das reicht beispielsweise von Silja Puranens Sexualität und Mutterschaft thematisierenden Arbeiten bis zu Margarete Warths ironischen Kommentierungen weiblicher Rollenklischees.

Für viele Arbeiten ist eine Eigenschaft von Filz bedeutungsvoll, auf die schon Robert Morris hingewiesen hat[1]. Für ihn hat Filz etwas Hautartiges, mit dem Körper in Beziehung Stehendes. Darauf kommt es auch vielen Künstlern an, die ihren Filz selbst herstellen und ihm dabei unterschiedliche Dichte und Qualität verleihen können, so daß er wie bei Hermine Gold einen Spannungsbogen von irdischer Schwere zu ätherischer Leichtigkeit zu schlagen vermag. Ganz direkt ist die Verbindung bei Célio Braga, dessen Thema der Körper in seiner Verletzlichkeit und Hinfälligkeit ist, wobei seine weichen Fragmente von Gliedmaßen und inneren Organen oft geradezu wie plastische Weiterführungen von Bildern Francis Bacons wirken. Auf die Welt des Organischen, gelegentlich wild Wuchernden verweist auch Hillevi Huse mit ihren vegetabilisch wirkenden Schlingformen, in denen das Problem der Anti-Form in veränderter Form wieder auftritt.

Einen Verweis auf den Körper enthält auch das Thema Kleidung, mit dem sich viele Künstlerinnen beschäftigen. Als die zweite Haut ist sie für Heather Belcher Ausgangspunkt ihrer künstlerischen Arbeit, bei der sie Kleidungsstücke in Bilder überführt. Andere Künstlerinnen halten an der Form des Kleides als einer Hülle für den Körper fest, lassen aber vielfach die Grenzen zwischen praktischem Gebrauch und freier künstlerischer Äußerung unbestimmt. Neben den tragbaren Kleidern – sei es für einen speziellen Anlaß oder in einer besonderen Stimmung – stehen die Phantasie- und Traumkleider, die eigentlich Objekt gewordene Märchen darstellen. So z.B. die als »Hülsen« bezeichneten Feenkleider von India Flint, bei denen in den Filz Flachs, Seide und andere Naturmaterialien eingearbeitet sind, oder die »Ghost Letters« von Jeanette Sendler, Kostüme einer Performance, die auf dem Friedhof Brompton in London stattgefunden hat. Leichter, bunter, das Thema der aus Mythos und Märchen bekannten Flügelschuhe ironisch brechend sind die »Schritt-für-Schritt-Stiefel« von Käthi Hoppler-Dinkel.

Mit diesen Schuhen wird bereits das Terrain tragbarer Kleidungsstücke betreten, das fast ausschließlich von »kunsthandwerklichen« Produkten beherrscht wird. Dabei handelt es sich um Kleider, die mit hohem handwerklichen Aufwand oft in Kombination mit anderen Stoffen, wie bei Christine Birkle, Charlotte Buch und Katharina Thomas, hergestellt werden. In jedem Fall sind es sehr individuelle Kleidungsstücke, die sich, wie beispielsweise die Creationen von Heidi Greb und Cristina Fröhlich, von aktuellen modischen Trends weitgehend fernhalten. Zwar haben sich auch führende Modedesigner wie Issey Miyake auf das Material Filz eingelassen, in großem Maßstab durchgesetzt hat es sich aber nicht. Es sind eher Außenseiter wie die finnische Designerin Anna-Kaarina Kalliokoski, die mit einer kompletten Filz-Kollektion aufwarten.

Auf dem Gebiet modischer Accessoires ist das Angebot wesentlich größer, doch sind auch hier individuell gestaltete kunsthandwerkliche Arbeiten vorherrschend, die poetisch verspielt wie bei Heyke Manthey oder von überraschendem Witz wie bei Ingrid Thallinger sein können. Als Designerin hat sich Bernadette Ehmanns mit Filztaschen in klaren geometrischen Formen hervorgetan. Ihr an die Seiten stellen ließen sich die Ringe von Kay Eppi Nölke und Bernhard Früh, die aus Filz sehr strenge Gebilde gestaltet haben, denen spielerischer Witz nicht fremd ist, wobei Bernhard Früh sich aber auf sehr kleine Auflagen beschränkt. Beliebt sind ansonsten Hüte und Mützen, die auch als Verkleide-Utensilien dienen können, z.B. von Jorie Johnson, Annette Stoll und Jeanette Sendler.

Den Filz für sich entdeckt haben auch einige jüngere Möbeldesigner, die dieses Material bei Stühlen und Sofas nicht einfach als Bezugsstoff verwenden, sondern direkt als Sitzfläche, die sich den Körperformen und Bewegungen des Benutzers anpaßt. Am konsequentesten hat das der finnische Designer Ilkka Suppanen mit seinem »fliegenden Teppich« verwirklicht, einem Sofa, das nur aus einer großen auf einem dünnen Stahlgestänge geradezu schwebenden Filzmatte besteht. Die Flexibilität des Materials nutzt auf andere Weise Reto Kaufmann, dessen aus einem Filzstreifen zusammengerollter Hocker ausgerollt auch als Liege verwendbar ist.

In den Jurten asiatischer Nomadenvölker spielten Teppiche eine wichtige Rolle. An den Mustern läßt sich ihre Herkunft leicht bestimmen. Wenn sich europäische Designer mit Teppichen beschäftigen, müssen sie eine eigene Formensprache finden, die sich von den traditionellen Mustern abhebt. Das wird schon dadurch erleichtert, daß die modernen Teppiche Industriefilz verwenden. In der Regel ist er homogen eingefärbt, und die Musterung geschieht oft mit einfachen technischen Mitteln, etwa indem Löcher in die Fläche gestanzt und diese u.U. mit einem anderen Material unterlegt werden. Die Dekoration ist weitgehend zurückgedrängt, sehr sparsam, die große, geschlossene Fläche und die Haptik des Materials werden betont.

So scheint Filz, der bei den nomadisierenden Volksstämmen Asiens jahrtausendelang das bevorzugte Material für Wohnstätten (Jurten) und deren Innenausstattung mit Teppichen war, auch in europäische Wohnungen Eingang zu finden, auch wenn er hier nicht dieselbe lebenserhaltende Bedeutung wie in seinen Ursprungsländern hat. Sie ist in den Entwürfen zeitgenössischer Designer nicht mehr erkennbar, die ihn aus anderen Gründen, seiner Flexibilität, Homogenität und Haltbarkeit wegen schätzen. Dem Ursprung näher kommen viele Künstlerinnen, für die die Filzherstellung ein Akt bewahrender Pflege ist, die sie einem Stoff zuteil werden lassen, mit dem sich wie mit kaum einem anderen die Vorstellung von Schutz und Wärme verbindet. Sie war schon für Joseph Beuys der Anlaß, sich als Künstler mit Filz zu beschäftigen. Darüber blieben andere Eigenschaften von Filz, »seine feste, verdichtete Struktur und seine meist dumpfe Farbigkeit«, durch die er »auch etwas Undurchdringliches, mit lebloser Materie Assoziiertes«[2] besitzt, für ihn bedeutungslos. Auch von den meisten Künstlerinnen wird Filz nicht als »ein autistischer Stoff«[2] empfunden. Sie haben ihm diese Eigenschaft auch gründlich ausgetrieben, indem sie Filz eine reiche Farbigkeit und Vielfalt der Strukturen bis hin zur Transparenz verliehen haben. Auf diese Weise ist Filz ein moderner Werkstoff geworden, in dem dennoch eine uralte Geschichte lebendig ist.

1 Interview mit Phil Patton, in: Art News 82, Nr. 10 (Dez. 1983), S. 50.
2 Wim Beeren in: Robert Morris. Recent Felt Pieces and Drawings. Hannover 1997, S. 19.

**Comme des Garçons,
Herbst/Winter 1995/9**
Comme des Garçons, c
autumn/winter 1995/96

Felt - Art, Crafts, Design

Peter Schmitt

Felt is 'in'. One might be tempted to ask whether the rediscovery of a material which long before Joseph Beuys declared it a symbol of warmth, life, cosiness and security has anything to do with the increasing coldness that is seeping into our social fabric. Be that as it may, felt has aroused fresh interest in artists, particularly women artists, worldwide since the late 1980s. It is not so much the industrially manufactured product, with which two artists, Joseph Beuys and Robert Morris, were already working, each in his entirely distinctive way, as often the process of felting itself which has moulded the creative statements made by artists using it.

The younger generation of artists working with felt still regards Joseph Beuys as a sort of father figure. He lent felt, like other materials which played a role in his work, symbolic meaning and pointed out its origins from nomadic culture, which was central to his way of thinking. Even though, according to his ideas of education, it was not Beuys' intention to find direct successors, he unmistakably broke the ground for the discovery of felt as a creative medium. Ines Tartler makes a direct reference to him with her felt suit 'alles wollen nach innen schlagen' (to turn all wanting inwards) and her felt apron with underpants, metaphorical clothing which provokes palpitation yet, as an autonomous object, questions assumptions.

The influence exerted by Robert Morris does not extend so much to the uses to which felt can be put, although his 'felt pieces' represent an important complex within his work as a whole, as to the concept of Anti-Form, which he, importantly, demonstrated with this material. He was interested in what shapes this heavy yet flexible material might form under the effect of the law of gravity and other factors when he cut into a roll of felt and then fastened it to the wall with hooks. The artist was only presenting a few of many possible variables yet he eschewed directing the process of shaping. In thus dismissing the traditional role of the artist as painter or sculptor, he was performing a paradigm change that was to have far-reaching effects.

The original materiality of felt, with which Morris was still working, has been for the most part suspended in Ernst Köster's work. Köster has removed the flexibility of felt by gluing several plates of it together, thus turning it into a rigid material from which he saws out his technoid shapes. Nor is the use of felt instantly recognisable in Ulrike Hein's work. She reproduces modern weapons of war in felt with deceptive authenticity, translating their inherent qualities of rigidity and deadliness into softness and warmth without depriving the objects of what makes them so terrifying.

It is their specific material properties which lend the wall and floor works, reductive as they are in respect of form and colour, by Maisa Tikkanen and Verena Gloor their distinctive character. The two artists work in influences from Minimal Art, although they subvert its forbidding severity by using such a warm material and by making visible the process of crafting. Their objects not only occupy space; they give it a meditative centre. Gunilla Paetau Sjöberg's pieces, with their flickering blacks, reveal a similar quality. Ingeborg Verres, on the other hand, unites amorphous soft felt in her 'Lichtfeld' (Lightfield) with rigid glass tubing glowing from within in a sharp contrast to create an artificial landscape which is mysteriously quiet.

More than in the sculptural qualities of felt, contemporary women artists seem to be interested in the possibilities of felt as a support for pictures. For the series she calls 'Belichtungszeiten' (Exposure times) Ute Lindner has taken this function literally. She uses red felt stretched across walls which was once the backdrop for pictures hung in the Kassel Gemäldegalerie. Exposed to light over a long time, these rolls of felt have almost entirely lost their original colour. All that shows up on the faded ground are the outlines of the pictures which once hung on it, darker surfaces, the shadows of pictures which have now become pictures themselves.

Felt is the support for pictures in the direct sense of the word in Silja Puranen's work, which deals in highly personal formulations with the role and fate of women. Transferring photos from her own life on to felt, she links them with other 'female' materials such as lace. In referring to her work, Denise Bettelyoun talks of 'felt panel pictures'; her representations of animals, which she felts into the natural white support, are not merely coincidental references to Joseph Beuys' drawings. After all, they, too, are distinguished by references to felt as a material.

Some women artists develop their pictures directly from the process of felting. As a result, the coloured surfaces and motifs are not sewn together; they are felted together. Examples of this are the carpets and wall hangings made by Inge Evers, May Jacobsen Hvistendahl and Turid Kvalvåg as well as Claudia Merx's double-sided felt paintings, which, although their patterns are so abstract, are still recognisably utility textiles.

The prerequisite for these works, what has made them possible, is the rediscovery of the craft of felting by hand, which has remained very much alive among the nomadic peoples of Central Asia. This renewal of interest in felting owes a great debt to the ethnological publications of Mary E. Burkett. Of almost equal importance are the efforts made by Mari Nagy and István Vidák, whose point of departure is Hungarian folk art and whose investigations have taken them as far afield as Turkey and Central Asia. Their activities have not been limited to collecting and recording; they have also, skilled as they are in crafts, tried out the techniques of felting themselves. Although they have retained motifs traditional in folk art, as teachers they have succeeded in exerting an influence which has reached far beyond their own country. Adopting Inge Evers' motto, 'felt the world together', a community has grown up around the workshops in Kecskemét for which the process of felting has been of paramount importance because, as the statements made by numerous participants prove, it unites craftswoman-like creativity with the meditative aspect.

Almost all the artists in this field are women and it represents an enclave which has little contact with the official contemporary art scene. Felt has this quality in common with other forms of creative expression, most of them linked with materials used in the applied rather than the fine arts. Moreover, felt seems to have an inherent quality which causes many women artists to recognize in it a medium which suitably reflects their situation. This perception runs the gamut from Silja Puranen's work dealing with sexuality and motherhood to Margarete Warth's ironical commentaries on feminine gender stereotyping.

For so much of this work one quality inherent in felt is of paramount importance, a property which Robert Morris[1] was among the first to point out. To him felt has something skin-like about it; it has a relationship with the body. This is what matters to so many of these artists, who make their felt themselves so that they can have it in various weights, thicknesses and qualities. In Hermine Gold's work felt ranges from earthy heaviness to ethereal lightness. Célio Braga's link is direct; his theme is the human body in all its vulnerability and fragility; his soft fragments of limbs and organs often look like nothing so much as Francis Bacon's paintings continued on into sculpture. Hillevi Huse also points to the realm of the organic with vegetal shapes that seem to be looped into each other and occasionally to grow like lush vegetation, representing a re-emergence of the problem of Anti-Form.

The theme of clothing, dealt with by so many women artists working in felt, also encompasses a reference to the human body. As a second skin, Heather Belcher has taken it as her starting-point in her work, translating clothing into pictures. Other women artists retain the shape of dresses as an envelope for the body yet leave the boundaries between practical uses and free aesthetic statements blurred. In addition to wearable clothes – whether worn for special occasions or when one is in a particular mood – there are fantasy and dream clothes, which actually represent the object turned into a fairy tale. Thus the as 'Hülsen' [husks] fairy dresses made by India Flint, in which flax, silk and other materials are worked into the felt, or Jeanette Sendler's 'Ghost Letters', costumes for a performance staged at Brompton Cemetery in London. Käthi Hoppler-Dinkel's 'Schritt-für-Schritt-Stiefel' [Step-by-Step Boots] are lighter, more colourful, thematically derived from the winged shoes of myth and fairy tale.

With these shoes we have entered the territory of wearable clothing, which is almost exclusively dominated by 'crafts' products. These are individual works which are created, often with great expenditure of craftswomanly skills in combination with other materials, by artists like Christine Birkle, Charlotte Buch and Katharina Thomas. In any case, these are often highly distinctive wearables which, as the creations of Heidi Greb and Cristina Fröhlich show, are at a far remove from current trends in fashion. Leading fashion designers like Issey Miyake have, of course, ventured to use felt as a material but it has not prevailed widely. Most of the designers who present complete collections of felt clothes are on the fashion fringe like the Finnish designer Anna-Kaarina Kalliokoski.

In the field of fashion accessories, the range on offer is much broader yet here, too, the individual crafts character predominates; it can be poetic and playful like Heyke Manthey's work or astonishingly witty like Ingrid Thallinger's. Bernadette Ehmanns has emerged as a designer to reckon with: her stunning felt bags are conceived in terms of clear geometric form. Similarly, the rings made by Kay Eppi Nölke and Bernhard Früh, who have shaped astringent forms from felt, are not without playfulness and are indeed quite witty. Bernhard Früh, however, restricts himself to very limited editions. Other popular creations in felt are hats and caps, which can also be used as vehicles for disguise. Jorie Johnson, Annette Stoll and Jeanette Sendler spring to mind in this connection.

A young generation of furniture designers has also discovered felt as a material for chairs and sofas but not merely as a covering; they are using it directly to make seats which adapt to the shapes and movements of the body. The most consistent realisation is the 'Flying Carpet' invented by the Finnish designer Ilkka Suppanen, a sofa which consists in nothing but a felt mat which seems to hover over a framework of thin steel rods. Reto Kaufmann has taken full advantage of the flexibility of felt as a material in an entirely different way. His pouffe (hassock), made of rolled-up strips of felt, unwinds for use as a lounge.

Carpets have always played a major role in the yurts of Asian nomadic peoples. The patterns on these carpets make it easy to identify their origins. When European designers become involved with carpets, they have to find a formal idiom of their own, which differs noticeably from traditional patterns. This is facilitated by the circumstance that modern carpeting is made of industrial felt. It is usually dyed in one colour and the designs on it have been made with simple technical means: holes may have been stamped in it and then backed with other materials. Such decoration is usually reduced to essentials, emphasizing the large, closed surface and the tactile quality of the material itself.

The material of choice for thousands of years in the dwellings (yurts, which were laid out with carpets as their only furnishings) among the nomadic peoples of Asia, felt seems at last to have come into its own in European homes even though it does not have the same importance for survival here as it does in the distant lands it comes from. It is not always recognisable as such in the work of contemporary designers, who appreciate it for entirely different reasons: for its flexibility, homogeneity and durability. Many women artists have come closer to its origins. For them felting is an act of caring and nurturing, which they lavish on a material which, like no other, unites the concepts of protection and warmth. That was the reason why an artist like Joseph Beuys chose to work with felt. Other inherent properties of felt, 'its firm, compacted structure and its usually subdued colour', which lend it 'also something impenetrable, associated with inanimate material'[2], were meaningless as far as he was concerned. Even most women artists working with it do not regard felt as 'an autistic material'.[2] They have expunged this quality from it by lending it rich colour and a great variety of textures ranging all the way to transparency. Thus felt has become a thoroughly modern material to work with yet its ancient traditions are indeed very much alive.

1 (…) felt has anatomical associations; it relates to the body – it's skinlike. The way it takes form, with gravity, stress, balance, and the kinaesthetic sense, I liked all that.' – Interview with Phil Patton, in: Art News 82, no. 10 (Dec. 1983), p. 50.
2 Wim Beeren in: Robert Morris. Recent Felt Pieces and Drawings. Hannover 1997, p. 19.

Vom Reichtum eines armen Materials

Filz bei Joseph Beuys

Kirsten Claudia Voigt

Prominenter als 1968 im Van Abbemuseum in Eindhoven war Filz noch nie exponiert worden: »Links vom Haupteingang hing eine große weiche Plastik von Morris schlaff von der Wand herunter, so daß es schwierig war, ihre Form zu analysieren. Es sah so aus, als hätte man ein Quadrat aus dickem grauem Filz mit einer geradlinigen Spirale geschnitten. Rechts hing ein Beuys von ähnlichem Format und Material, das wie die abgezogene Haut eines großen Tieres – zu klein für einen Elefanten, also vielleicht ein Rhinozeros – aussah. Es stellte sich heraus, daß es sich um seine inzwischen berühmte Filzdecke für einen Flügel handelte, wobei jedes klauenartige Bein sorgfältig ausgearbeitet war. Durch diese Werke offenbarten sich beide Künstlerpersönlichkeiten. Der Amerikaner: kühl, nachdenklich, abstrakt, minimal. Der Deutsche: weißglühend, visionär, geheimnisvoll, vielschichtig.«[1]

Der britische Pop-Art-Künstler Richard Hamilton, dessen Beschreibung hier zitiert ist, begriff die Ausstellung in Eindhoven als eine Schlüsselschau, die den Kontrast zwischen der Entwicklung der amerikanischen und europäischen Kunst nach 1945 dokumentierte. Der Minimalist und der »Mystiker«, der Formalist und der Idealist begegneten einander auf demselben Terrain mit demselben Material.

Außer Frage steht mittlerweile, daß Joseph Beuys Filz zuerst, 1963/64, systematisch für sich entdeckte und Robert Morris dieses Material erst ab 1967 benutzte. Das Eindhovener Zusammentreffen verlief denn auch nicht irritationsfrei. Beuys zeigte sich befremdet von Morris' Auftritt mit Filz und empfand ihn als epigonal.[2]

Welche Absichten Beuys mit seinem Material verfolgte, ist längst kunsthistorisches Allgemeingut. Das kann nicht verwundern, wenn man sich daran erinnert, daß neben dem Hut (aus Filz) vor allem Filz und Fett zu seinen öffentlich immer wieder benannten, oft verhöhnten Markenzeichen wurden. Mag die Verwendung von Filz für die künstlerische Arbeit auch durch biographische Begebenheiten nahegelegt worden sein, Beuys' nachhaltige Beschäftigung mit diesem Material verdankt sich nicht etwa einer singulären Qualität des Stoffes oder einer einzigen bildnerischen Intention, sondern grundlegender der fortgesetzten Suche des Künstlers nach Erweiterungen und seinem Drang zur präzisen Erforschung stofflicher Phänomene. Filz wird in seinem Œuvre deshalb aus mindestens fünf Perspektiven reflektiert und instrumentalisiert.

Als plastischer Stoff kann Filz formal vor allem in zwei Zuständen, als Schichtung oder Umhüllung, eingesetzt werden. Beuys' Interesse an der Arbeit mit planen Materialien – dünnen Schiefertafeln, Kupferplatten, Zeitungen, die gestapelt werden – reicht bis zu den Ursprüngen seines künstlerischen Schaffens zurück und erklärt sich aus seiner Affinität zum Relief einerseits und zu ans Geologische erinnernden Schichtungen andererseits. Außerdem ist er als Bildhauer lebenslänglich auf der Suche nach der Plastik als Gefäß. Hüllen, Schläuche, Höhlen, Räume faszinieren ihn nachhaltiger als hermetische, plastische Konstellationen. Deshalb ist seine monumentalste und eindrucksvollste Filzplastik denn auch ein großes, Geborgenheit vermittelndes Environment: »Plight«, eine riesige Akkumulation aus gestapelten Filzsäulen, öffnet sich als warmer Raum, in dem Schritte leise klingen, in dem ein stummer Flügel und ein wohltemperiertes Thermometer aufbewahrt werden.

Die Materialfarbe **kam Beuys' theoretischen Absichten entgegen. Als »Farb«-Stoff barg der graue Filz (virtuell) das gesamte Spektrum, was Beuys didaktisch nutzte, wenn er seine Gegenbildtheorie zu erklären suchte: »Ja, der Beuys arbeitet mit Filz, warum arbeitet er nicht mit Farbe? Aber die Leute denken nie soweit, daß sie sagen: Ja, wenn er mit Filz arbeitet, könnte er nicht vielleicht dadurch in uns eine farbige Welt provozieren? Es gibt ja bekanntlich das Phänomen der Komplementarität (...). Die Leute sind sehr kurzsichtig mit dieser Argumentation, wenn sie sagen: der Beuys macht alles so mit Filz, und dann will er etwas aussagen von KZ. Ob ich nicht daran interessiert bin, durch diese Filzelemente die ganze farbige Welt als Gegenbild im Menschen zu erzeugen, danach fragt keiner. Also: eine lichte Welt, eine klare lichte, unter Umständen eine übersinnlich geistige Welt damit sozusagen zu provozieren, durch eine Sache, die ganz anders aussieht, eben durch ein Gegenbild.«[3]**

Die nicht optisch wahrnehmbaren physikalischen Eigenschaften **von Filz, Qualitäten, um die sich die Bildhauerei konventionellerweise nicht kümmerte, rückten außerdem ins Zentrum des gestalterischen Interesses. Die von Beuys angestrebte Erweiterung des Begriffs von Plastik bedingte es, bei der Materialwahl die Potentiale, mögliche Funktionen und Funktionsweisen, der Stoffe mitzubedenken. Am Filz ließ sich dessen Absorptionsfähigkeit, seine Dämmqualität reflektieren. Diese Faktoren wurden relevant, weil es dem Künstler um die Veranschaulichung von Abstrakta wie Energien und Aggregatzuständen ging, die ihrerseits metaphorisch für soziale Zusammenhänge standen.**

Der Kontext, **aus dem Filz ins Werk einging, war kein beliebig indivi-dueller. Mit der Wahl von Filz verbindet sich ein Rekurs auf Militär, Krieg und Not. Beuys erwähnt selbst den Bezug zu Konzentrations-lagern. Damit werden triftige historische Kontexte zum Bestandteil der Werke.**

Auch die Materialgenese **thematisierte Beuys. Ausgangsstoff sind Tier-haare, die durch Kompression – einen plastischen Prozeß – zu einer Form verdichtet werden, die jedoch in ihrer Binnenstruktur chaotisch bleibt. Damit sind dem Filz jene beiden polaren Prinzipien inhärent, die für Beuys Eckpfeiler seiner Theorie von Plastik wurden: Chaos und Ordnung, zwischen denen ein bewegter Formprozeß vermittelt.**

Die derzeit beobachtbare modische Rückkehr von Filz in den Alltag hätte Beuys gewiß mit einigem Amüsement beobachtet. Schließlich war es ihm vor allem auch darum gegangen – und gelungen –, den Reichtum eines »armen« Materials begreiflich zu machen.

1 Richard Hamilton, in: Beuys zu Ehren. Hrsg. v. Armin Zweite. München 1986, S. 356.

2 Beuys erklärte in einem Interview: » ... ich hätte an Morris' Stelle das nicht gemacht. Wenn ich sehe, daß ein Mensch etwas macht, was schon durch die formalen Mittel so extrem liegt, dann würde ich es nicht machen, auch wenn ich es gerne machen möch-te, dann würde ich sagen, gut er hat es gemacht, es ist gemacht, dann würde ich versuchen, etwas anderes zu machen oder gar nichts zu machen.« Zitiert nach Dirk Luckow: Joseph Beuys und die amerikanische Anti Form-Kunst. Berlin 1996, S. 80.

3 Joseph Beuys, in: Jörg Schellmann und Bernd Klüser (Hrsg.): Joseph Beuys, Multiples. Werkverzeichnis Multiples und Druckgraphik 1965–1985. München 1985 (6. Aufl.), o. S.

Joseph Beuys
»Plight«, Filzstrukturlandschaft,
Oktober 1985
'Plight', structured felt landscape,
October 1985

Joseph Beuys
»Infiltration homogen für Konzert-
flügel«, 1966, Centre Pompidou, Paris
'Infiltration Homogenous for a Concert
Grand', 1966, Centre Pompidou, Paris

Joseph Beuys
»Filzanzug«, 1970, Filz, genäht und
gestempelt, ca. 170 x 60 cm
Sammlung Froehlich, Stuttgart
'Felt Suit', 1970, sewed and stamped,
ca. 170 x 60 cm
Froehlich Collection, Stuttgart

On the Richness of a Poor Material

Felt by Joseph Beuys

Kirsten Claudia Voigt

Felt has never been exhibited more prominently than in the Van Abbemuseum in Eindhoven in 1968: 'To the left of the main entrance a large, soft sculpture by Morris hung down so limply from the wall that it was difficult to analyze its form. It looked as if a square had been cut out of thick grey felt with a straight spiral. To the right hung a Beuys of similar dimensions and material which looked like the hide skinned from a large animal – too small for an elephant, perhaps, then, a rhinoceros. It turned out to be what has since become a famous felt cover for a grand piano with each claw-like leg meticulously worked out. Two artists are revealed by these works as two distinct personalities. The American: cool, contemplative, abstract, minimalist. The German: white-hot, visionary, mysterious, sophisticated.'[1]

The British Pop artist Richard Hamilton, whose description is paraphrased above, thought of the exhibition in Eindhoven as a key show which documented the development of American and European art after 1945. The Minimalist and the 'mystic', the formalist and the idealist met on equal terms using the same material.

Since then it has been clear that Joseph Beuys was the first to discover felt systematically and that Robert Morris did not start using this material until 1967. The Eindhoven encounter was not without hitches. Beuys did not care for Morris' performance with felt and looked down on him as an epigone.[2] What Beuys intended to do with his material has long since made history. That is not surprising when one recalls that, apart from his hat (a felt hat), felt and fat had become his public trademark, one that was frequently mentioned and often derided. Although the use of felt for his creative work may also have seemed obvious for biographical reasons, Beuys' long-term preoccupation with this material is not due to any distinctive quality it might have or to a single artistic intention. Instead, a more basic underlying reason was the artist's continuing search for enlargement and his urge to investigate material phenomena. Felt is, therefore, studied and used from at least five different angles in his œuvre.

As a plastic material felt can be used in the formal sense in two states, as a stack or envelope. Beuys' interest in working with flat materials – thin slate tiles, copper plates, piled up newspapers– goes all the way back to the well-spring of his creativity as an artist and the explanation for this is his affinity for the relief on the one hand and, on the other, for layers recalling geological strata. Moreover, as a sculptor, he sought all his life for sculpture as a vessel. Envelopes, tubes, caves, spaces had a more enduring fascination for him than hermetic plastic constellations. For this reason his most monumental and impressive felt sculpture is also a spacious environment radiating security and cosiness: 'Plight', a vast accumulation of piled up felt columns, opens up as a warm room in which steps sound muffled, in which a mute grand piano and a well-tempered thermometer are kept.

The **colour of the material** matched Beuys' theoretical intentions. As a 'colour' material grey felt concealed (virtually) the entire spectrum, which Beuys used didactically when he tried to explain his theory of complementary colour: 'Yes, Beuys works with felt, why doesn't he work with colour? But people never think far enough to say: Yes, if he works with felt, couldn't he perhaps provoke a coloured world in us through it? There exists, as is well known, the phenomenon of complementarity (...). People are very short-sighted with this line of reasoning when they say: Beuys does everything just like that with felt and then he wants to make a statement about concentration camps. No one asks about whether I might not be interested in generating the entire coloured world as a complementary image in people. That is: so to speak, to provoke a light world, a clear, light world, possibly a supernatural spiritual world through a thing that looks entirely different, a complementary image, to be precise.'[3]

In addition, the **physical properties** of felt which cannot be perceived with the naked eye, qualities which conventional sculpture has disregarded, moved to the centre of formal and structural interest. Enlarging the concept of sculpture as Beuys sought to do required also taking into account the potential, possible functions and ways of functioning of materials. Thought should be given to the ability of felt to absorb, to insulate and mute. These factors became relevant because the artist was concerned with illustrating abstract qualities like energies and states which, in turn, stood metaphorically for social correlation. **The context** from which felt entered his work was not an arbitrary, individual one. The choice of felt implied returning to the military, war and need. Beuys himself mentioned the relationship to concentration camps. Thus apt historical contexts became part of the works. Beuys also made a point of the **genesis of his material.** Animal hairs which were compressed into a shape by being compacted – a sculptural process – yet remained chaotic in their inner structure were the original material. That means that the two polar principles are inherent in felt which became for Beuys the twin pillars of his theory of sculpture: chaos and order between which an eventful form process mediates.

It certainly would have amused Beuys to observe the current return of felt to fashion in everyday living. After all, what was also a matter of primary concern to him was making the richness of a 'poor' material understandable – and he certainly succeeded in doing so.

1 Richard Hamilton, in: Beuys zu Ehren. Ed. Armin Zweite, Munich 1986, p. 356.
2 Beuys made the following statement in an interview: '... I wouldn't have done that if I were Morris. If I see somebody doing something which is so extreme simply because of the formal means, I wouldn't do it, even if I wanted to do it; then I'd say, good, he did it, it's been done, then I'd try to do something else or do nothing at all.' Quoted in German in Dirk Luckow: Joseph Beuys und die amerikanische Anti Form-Kunst. Berlin 1996, p. 80.
3 Joseph Beuys, in: Jörg Schellmann and Bernd Klüser (eds.): Joseph Beuys, Multiples. Werkverzeichnis Multiples und Druckgraphik 1965 – 1985. Munich 1985 (6th impr.), n. p.

Dem Filz auf der Spur

Eine Liebesgeschichte zwischen Traum- und Raumzeitalter

Sabina Heck / Katharina Thomas

Geheimnisvolles fasziniert, Unerwartetes überrascht und solide Tatsachen überzeugen jeden, der sich auf die Erfolgsfährte eines Materials begibt, das in unseren Breitengraden wohl die längste Zeit ein Dasein als »Pantoffelheld« gefristet hat.

Josef Beuys machte ihn schon in den 60er Jahren zum Star seiner Installationen. Gegenwärtig erobert er die Laufstege und Designtempel internationaler Metropolen – wo Natürliches und die Vitalität solider Handwerkskunst mal wieder umschwärmt werden.

Sein archaisch-mystisches und gleichzeitig ultramodernes Eigenschaftsprofil verblüfft und bezaubert gleichermaßen.

Aus sagenumwobener nomadischer Vergangenheit kommend, bringt er alle Eigenschaften mit, die ihn zum Material des beginnenden dritten Jahrtausends werden lassen. Wir stellen vor: Der Filz.

Faser, Flocke, Vlies
Der Filz ist mehr als die Summe seiner Teile

Durch die Verwendung von Filz entsteht ein phantastisches Raumklima. Der Filz wirkt direkt auf seinen Betrachter, stärker noch auf jeden, der ihn auch nur einmal berührt. Der Filz ist eine visuelle, kinästhetische und emotionale Erfahrung, man kann ihn unmittelbar erleben und spüren.

Jeder Künstler, der professionell selbst Filz macht, ist davon überzeugt, daß bei dem oft Tage dauernden Entstehungsprozeß viel von der Haltung und Persönlichkeit des Machers mit einfließt; ein Hauch von Schutzzauber und Magie liegt in der Luft.

Tatsächlich ist Filz nicht nur nach praktischen und auch ökologischen Gesichtspunkten ein hochgeschätztes Material, auch seine unsichtbaren Fähigkeiten sind historisch dokumentiert: Als Tschingis Khan zum Mongolensturm rief, war Filz ein Stoff mit Magie. Filz schützte Krieger mit seiner Festigkeit und Dichte im Kampf. Römische Sklaven setzten sich nach ihrer Freilassung Filzkappen auf, als Symbol für ein Leben in Freiheit. Skorpione, Taranteln und Schlangen meiden Filz – ein Phänomen, das Nomaden damals und so manchem heute Reisenden ausgesprochen angenehme Dienste leistet.

Seine Entstehung: Ein glücklicher Zufall?

Um die Entstehung von Filz ranken sich Legenden: Als Gott alle Tierpaare der Erde unter Noahs Obhut stellte und Noah die Arche fertiggestellt hatte, überlegte er sich, wie er für Menschen und Tiere die ungewisse Reise angenehmer gestalten konnte, denn er sah die harten Holzplanken des Schiffes. Daraufhin bedeckte er den ganzen Boden mit Schafwolle. Nachdem nun Mensch und Tier vierzig Tage und Nächte in dieser Enge sitzend, liegend, stampfend, schwitzend die Sintflut überstanden hatten, war aus der lockeren und losen weichen Wolle ein zusammenhängendes Wollstück geworden. Noah hielt in seinen Händen einen festen aber dennoch flexiblen Stoff – Filz! Oder: In Persien stampfte Salomons Sohn wütend auf einen Wollvlies ein, weil es ihm nicht gelang, einen Teppich herzustellen, auch tränkte er dieses mit den bitteren Tränen der Verzweiflung. Als seine Wut vorüber war, entdeckte er, daß er auf wasserdichtem Filz stand.

Seine Geschichte: Schutz und Wärme auf dem Weg durch die Zeit

Vermutlich ist der Filz das älteste vom Menschen erzeugte Textil überhaupt. Historisch dokumentiert begleitet der Filz die Menschheit bereits seit den ersten Jahrhunderten vor Christus, sicherlich aber auch schon davor. 1929 fanden zwei russische Wissenschaftler bei den Ausgrabungen Filzgegenstände in den Grabfeldern von Pazyrik/Sibirien.

Das Altai-Volk pflegte prachtvolle Begräbnisse, die große Ähnlichkeit mit denen der nah verwandten Skythen hatten. Dinge des täglichen Lebens, wie Satteldecken, Wandbehänge, Applikationen auf Kleidung, Socken für Frauen und Männer, wurden aus diesem so überaus praktischen und vielseitigen Material ebenso angefertigt wie rituelle Gegenstände und Rangabzeichen.

Schutz und Schmuck bot der Filz den Nomaden der zentralasiatischen Steppen, aus denen er, wahrscheinlich über Ungarn, dann nach Europa gelangte. Die älteste Darstellung der Filzherstellung findet sich auf einer Wandmalerei an der Außenwand der Werkstatt von Verecundus in Pompeji. Produkte aus Filz kennt und fertigt man auch in den nordeuropäischen Ländern seit Menschengedenken. Von der Wiege bis zum Grab, als Stoff der Könige und der Armen, zieht er seinen Weg durch geographische Räume und Zeiten, ein steter Begleiter des Menschen.

Die erste Begegnung: Ein sinnliches Schlüsselerlebnis

Ich sehe Fasern als ein organisches Element, welches die organische Welt auf unserem Planeten bildet. Sind nicht alle lebenden Organismen aus Faserstrukturen gebildet? So auch der Mensch. (M. Abakanowicz)

1983 schlief ich zum ersten Mal in einer mongolischen Ger (Jurte). Ich erfuhr es als ein Eins-Sein mit mir und der Natur, eine absolute Ausgeglichenheit von Innen und Außen. Geborgen im Rund der Jurte und beschützt durch die besonderen Qualitäten des Filzes: Filz dämmt und speichert Wärme. Filz polstert und poliert. Filz nimmt, wie jede Wolle, bis zu 30 % Feuchtigkeit auf, ohne sich feucht anzufühlen. Filz ist lange Zeit wasserabweisend und winddicht. Auch heute leben die mongolischen Nomaden in ihren Ger, wie ich auf meinen verschiedenen Reisen durch die Mongolei erleben konnte.

Wohlbehagen mit Haut und Haar

Mit seinen Eigenschaften war und ist der Filz in vielen Bereichen einzusetzen, in denen der Mensch tätig ist und lebt.

So z.B: als Stoff für Mäntel, Jacken, Hüte, Schuhe, Taschen, Teppiche, Kissen; es gibt auch Schränke und Vorhänge aus Filz. Er eignet sich ebenso als biologisch gesundes Baumaterial, wie als eigenwilliger Stoff, aus dem Künstler ihre Träume und Visionen entstehen lassen. Seine Anziehungskraft und der Bann, in den er jeden schlägt, liegen jedoch anderswo: Filz ist ein Stück lebendige Natur, die ihren Charakter seit Tausenden von Jahren bewahrt hat. Er ist Wohlbehagen mit Haut und Haar. Man will und muß ihn berühren, an ihm riechen, sich auf ihn legen, mit allen Sinnen wahrnehmen und genießen. Dabei wird einem klar: ein Stück Wildheit und Ursprünglichkeit sind in ihm verborgen, die Erinnerung an Lagerfeuer, wilde Reiter und das Leben im Freien. Er verkörpert Muster, Lebensweisen und Traditionen, die in unserem kollektiven Gedächtnis gespeichert sind und deren Lebenslust und Kraft noch genauso wirken wie zu Zeiten der nomadischen Erfinder.

Seine Herstellung: Wasser, Wolle und harte Arbeit

Es ist nicht nur seine Eigenschaften, die jeden, der sich näher mit Filz einläßt und in seine Geschichte und Eigenart eintaucht, mit der ihm eigenen Ursprünglichkeit und Sinnlichkeit verbindet. Auch bei der Art der Herstellung folgt der Filzer einem uralten Rhythmus und Geschehen, dessen Grundelemente Wasser, Wolle und harte Arbeit sind, auf denen auch heute noch das Filzen basiert.

Stationen einer Verwandlung

Je nachdem, was man filzen möchte, verwendet man Wolle unterschiedlicher Schafrassen. Die Wolle ist kardiert (gekämmt) oder in der Flocke, möglichst noch mit etwas Wollfett belassen und erscheint entweder in ihren vielfältigen Naturfarben oder auch gefärbt.

Die Wollfasern werden exakt auf einer Schilfmatte nebeneinander gelegt. Die Anzahl der Lagen bestimmt die gewünschte Dicke des Filzes. So ist es möglich, Millimeter dünne Stoffe genauso wie Zentimeter dicke Teppiche herzustellen. Je dünner der Filz, um so exakter muß gearbeitet werden.

Alles wird mit heißem Wasser besprengt. Die Matte wird zusammen mit der feuchten, dampfenden Wolle um einen Innenkern gerollt und verschnürt, so daß ein Paket entsteht, ähnlich einem zusammengerollten Teppich.

Nun beginnt die Arbeit des rhythmischen Walkens. Je nach Größe der Rolle wird mit den Armen oder mit den Füßen, allein oder zu mehreren getreten (gewalkt). Der Prozeß ist immer der gleiche, doch entstehen mit wachsendem Können, Erfahrung und kulturellem Hintergrund unterschiedliche Handschriften.

Bewegung und stetiges Zusammenfügen gehören zum Filz.

Die Techniken haben sich verfeinert. Jeder Filzer hat seine eigenen Tricks, Geheimnisse, Rezepte und Arbeitsweisen. Längst hat der Filz den Sprung vom praktischen Alltagsgegenstand zum Träger einer künstlerischen Idee gemacht; international anerkannte Designer und Objektkünstler arbeiten mit diesem Material.

Inzwischen hat der Filz auch in die Lehrveranstaltungen der Hochschulen und Akademien Einzug gehalten.

Endgültig hat der Filz nun die Moderne erreicht. Wo er, als guter Nomade, natürlich nicht stehenbleiben, sondern seinen erfolgreichen Weg in die Zukunft fortsetzen wird.

Satteldecke aus Pazyryk im Altaigebirge, 400 v. Chr.
Saddle cloth from Pazyryk at the Altai Mountains, 400 BC

Mongolische Ger (Jurte) im Altaigebirge
Mongolian ger (yurt) at the Altai Mountains

**Innenraum einer mongolischen Ger (Jurte),
mit Filz bedeckt**
Interior of a Mongolian ger (yurt), covered with felt

Kasachischer Filzteppich
Kasakh felt carpet

Walken von Filz in der Mongolei
Fulling felt in Mongolia

**Mongolische Nomaden beim Einrollen von Wolle
zum Walken**
Mongolian nomads rolling up wool for fulling

Karen Page (USA)
'Stratum No 4', 1991

Katharina Thomas (D)
Abendkleid aus plissiertem
Filzstoff, 1999
Evening dress of pleated felt
material, 1999

Gert Pfafferoth/
Katharina Thomas (D)
»Expo Gastmahl«,Theaterproduk-
tion mit Filzstoffen, EXPO 2000,
Hannover
'Expo Gastmahl', theatre production
with felt materials, EXPO 2000,
Hanover

Tracking down Felt

A Love Story between Dream and Space Ages

Sabina Heck / Katharina Thomas

Mysterious things fascinate, the unexpected surprises and sound facts convince anyone who undertakes to follow up the success story of a material which, in our latitudes, has probably spent most of its existence 'under someone's thumb'.

By the 1960s Josef Beuys had made it the star of his installations. It is currently conquering the catwalks and design shrines of metropolises worldwide – wherever naturalness and the vitality of sound craftsmanship are again occupying centre-stage.

The mystique of its archaic yet ultramodern range of qualities is both astounding and enchanting.

It emerges from a legendary nomadic past, bringing with it all the properties which will make it the material of the early third millennium. Let us introduce it: felt.

Fibre, flock, fleece
Felt is more than the sum of its parts

Using felt creates a fantastic room climate. Felt has a direct effect on anyone looking at it and an even stronger one on those who touch it at least once. Felt is a visual, kinaesthetic and emotional experience; one can experience and feel its immediacy.

Every artist who makes felt professionally is convinced that the process of felting, which often takes days, puts a lot of his/her own attitudes and personality into it; a trace of apotropaic magic is left in the air.

In fact felt is not just a highly appreciated material, valued for practical and ecological reasons. Its invisible powers are also historically documented:

when Genghis Khan summoned up the Mongol hordes to ride to battle, felt
was a material with magic properties. With its firmness and density, felt protected warriors in battle. Those who had once been Roman slaves donned felt caps to show that they were freedmen as a symbol of a life henceforth to be led in freedom. Scorpions, tarantulas and snakes avoid felt – a phenomenon which has made felt so serviceable for nomads in the past and many a traveller nowadays.

How it came into being: serendipity?

The origins of felt are intertwined with the stuff of legend: when God placed pairs of all the earth's animals under Noah's protection, Noah gave some thought to making the journey into uncertainty more comfortable for man and beast since he realised how hard the bare planking of the ark's deck was. He decided to cover the entire deck with sheep's wool. After both people and animals had survived the deluge for all of forty days and nights in these cramped quarters, lying down, stamping about and sweating, what had been light, fluffy soft wool had become a single piece of firm yet flexible material – felt!

Or: In Persia Solomon's son stamped in range on a fleece of wool because he had not succeeded in making a carpet; he also wet it with bitter tears of desperation. By the time his rage was spent, he discovered that he was standing on waterproof felt.

Its history: protection and warmth down through time

Felt is probably the earliest textile made by man. Historically documented, felt accompanied humankind from the earliest centuries before Christ and is highly likely to have done so even before that. In 1929 two Russian archaeologists found felt objects in the necropolises of Pazyrik in Siberia.

The people of Altai practised elaborate burial customs and funeral rites which strongly resembled those of the Scythians, to whom they were closely related. Like ritual objects and badges of rank, objects used in daily life, like saddle blankets, wall hangings, appliqués on clothing, socks for both men and women, were made of this so thoroughly practical and versatile material.

Felt afforded the nomads of the Central Asian steppes protection and adorned them like jewellery. Thence it spread to Europe, probably via Hungary. The earliest representation of felting is in a wall painting on the outside of the Verecundus workshop in Pompeii. Felt products have been known and made in the northern European countries since time immemorial. From cradle to grave, a material for kings and commoners, felt has gone its way through geographical space and time, faithfully accompanying humankind.

The first encounter: a key experience in sensuousness

> *I see fiber as the basic element constructing the organic world*
> *on our planet. It is from fiber that all living organisms are built*
> *– the tissues of plants and ourselves. We are fibrous structures.*
> *Handling fiber, we handle mystery.* (M. Abakanowicz)

In 1983 I slept in a Mongolian ger (yurt) for the first time. I experienced it as being one with myself and nature, absolute balance both inside and outside. Cosily contained within the circle of the yurt and protected by the special properties of felt. Felt insulates and stores warmth. Felt upholsters and polishes. Felt absorbs, like all wool, up to 30% humidity without feeling damp. Felt is water-repellent for a long time and is impervious to wind. Even today Mongolian nomads live in gers, as I have seen on my various journeys through Mongolia.

Totally cosy comfort

Its particular properties have always made felt useful in many areas in which people live and are actively doing something. For instance it is used as a material for making coats, jackets, hats, shoes and bags, carpets and cushions; there are even wardrobes and curtains made of felt. It is also just as suited for use as a healthful organic building material as it is a material with a will of its own from which artists can bring forth their dreams and visions. Its appeal and the spell it casts, however, are on a different plane: felt is a piece of living nature which has retained its true character for thousands of years. It represents total cosy comfort. You want to touch it, you have to touch it, sniff it, lie down on it, take it in with all your senses and simply enjoy it. In so doing, you become aware that something wild and elemental is concealed in it, the memory of camp-fires, wild riders and outdoor living. It embodies patterns, ways of life and traditions which are stored in our collective memory and whose zest for living and powers are still as strong as when the nomads first invented it.

Making it: water, wool and hard work

Its properties are not all that attach everyone who lets him/herself in for closer contact to felt and its history and distinctiveness, with its unique primal and sensuous qualities. Also the way felters follow a primal rhythm and rite in making felt, whose basic elements are water, wool and hard work, elements on which felting is still based today.

Stages of a transformation

Depending on what one wants to felt, one selects a particular wool from one or many breeds of sheep. The wool is carded (combed) or left as flock, if possible retaining some of its natural lanolin, and is used either in any of its many natural colours or is dyed.

The wool fibres are laid down precisely next to each other on a reed mat. The number of layers depends on the thickness of felt desired. Consequently it is just as possible to make materials that are only millimetres thick as it is to produce carpets centimetres thick. The thinner the felt, the more precision called for in working it.

The whole thing is sprinkled with hot water. The mat is rolled, together with the damp, steaming wool, round an inner core and tied up to form a roll resembling a rolled-up carpet. Now begins the rhythmic work of fulling. Depending on how big the roll is, it is beaten with the arms or trodden with the feet of one person alone or several together (fulling). The fulling process never varies yet, the more skilful and experienced the felter and depending on his/her cultural background, the signatures that emerge are diverse. Movement and continual piecing together are part and parcel of felt. The techniques of felting have become more sophisticated. Every felter has his/her own tricks, secrets, recipes and methods of working. Felt has long since made the leap from being a practical everyday object to carrying an aesthetic idea; designers and object artists known world-wide work with this material.

By now felt has even become part of the curricula of universities and academies.

Felt has become definitively modern. As a real nomad, however, it naturally will not just settle there in one place but will continue its successful journey on into the future.

»Ghost Letters«
Kostüme für eine Performance
auf dem Friedhof Brompton
Wolle, gefilzt, Flechten, Gras,
Papier, Schnur, Garn
August 1999
»Ghost Letters«
Performance art costumes shown
at Brompton Cemetery
Wool felted, lichen, grass, paper,
string, yarn
August 1999

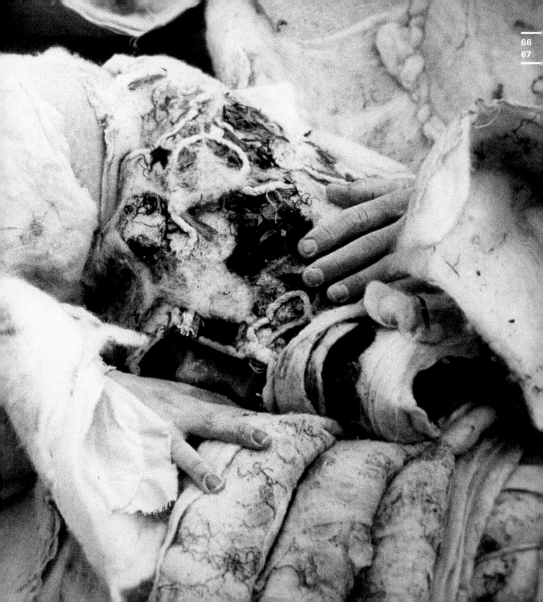

Die Kostüme gehören zu einem Konzept, das Kostümkunst als eigenständige Kunstform versteht. Sie stammen aus der poetischen Performance mit dem Titel »Ghost Letters«, die 1999 auf dem Friedhof Brompton in London stattfand. Sie soll im Jahr 2001 auch in Berlin und auf drei weiteren europäischen Friedhöfen aufgeführt werden.

The costumes represent Fibre Art as an art form in its own right. They stem form 'Ghost Letters', a poetry performance which was shown at Brompton Cemetery in London in 1999. In 2001 »Ghost Letters« will be shown in Berlin and in other three European cemeteries.

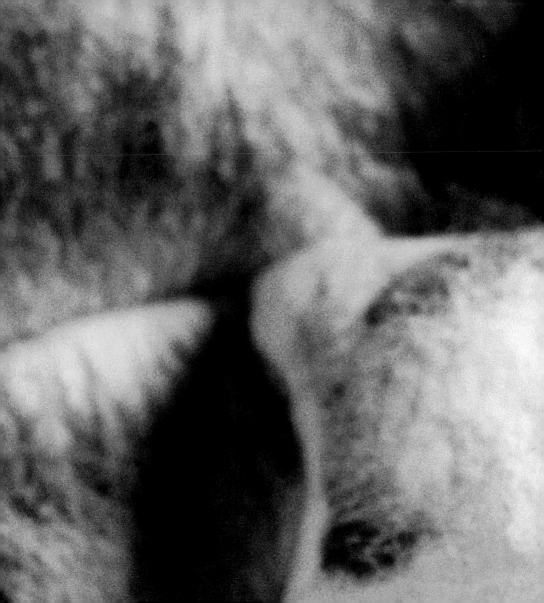

Hüte
Wolle, gefilzt
1999
Hats
Wool felted
1999

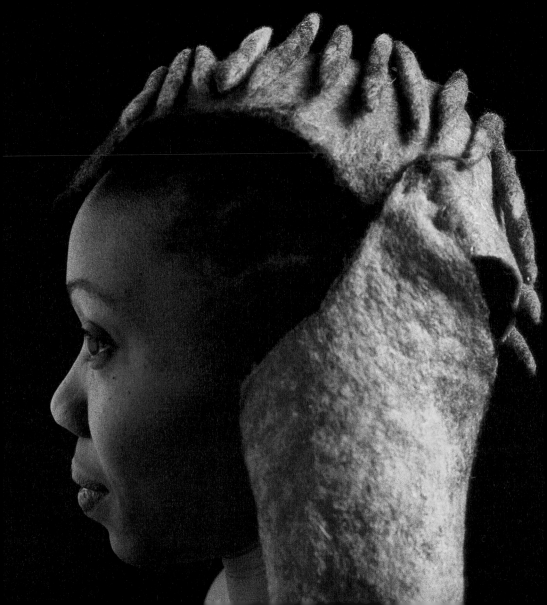

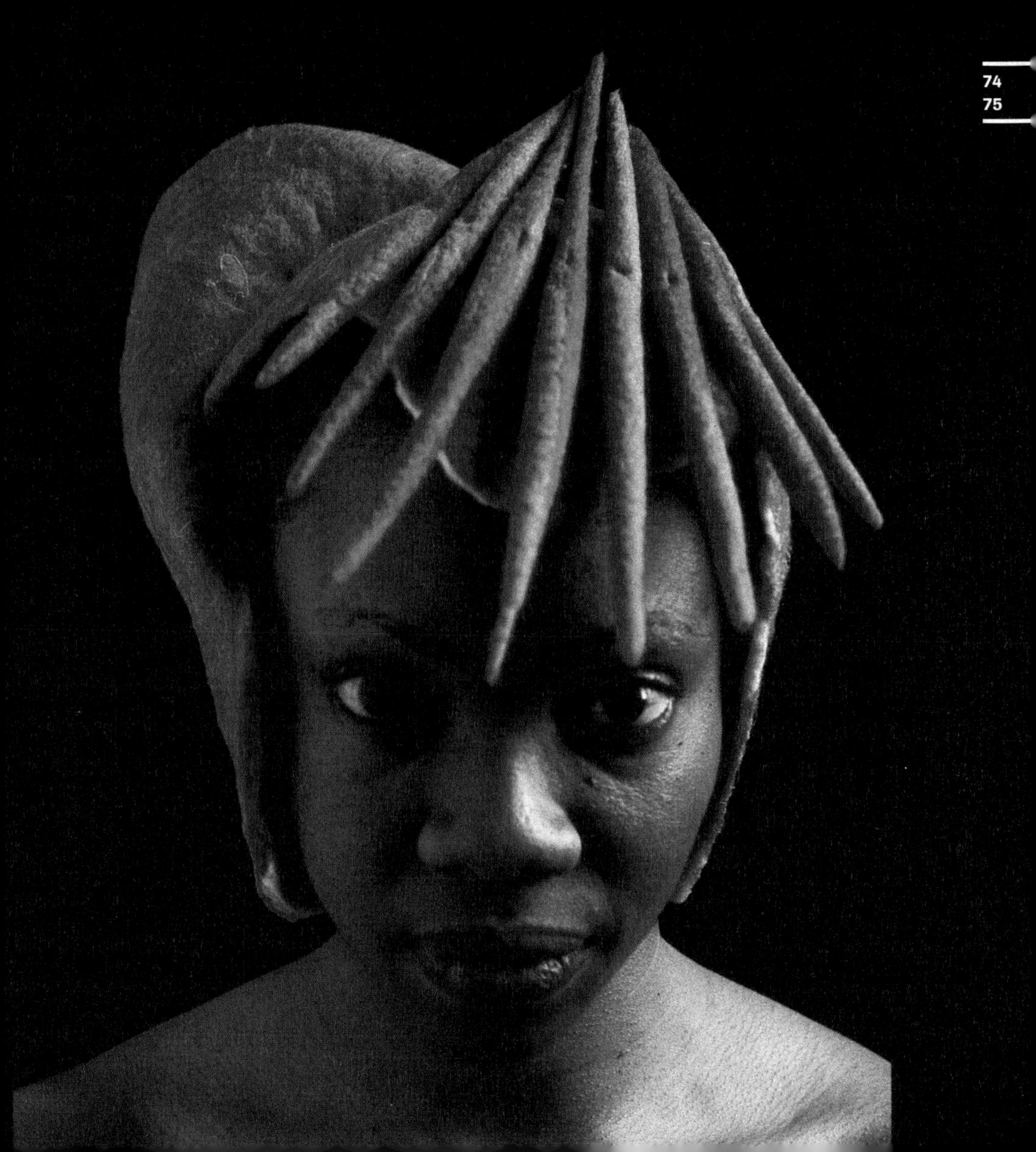

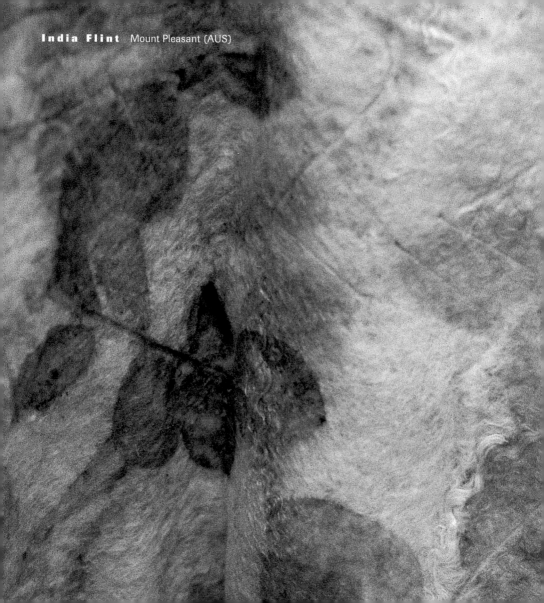

India Flint Mount Pleasant (AUS)

India Flint bedient sich zum Färben ihrer Textilien eines selbstentwickelten nicht-chemischen Verfahrens mit pflanzlichen Materialien, das sie als Eco-Print bezeichnet. Über ihre Arbeit schreibt sie: »Durch Zufall kam ich in Australien zur Welt. Als Kind hörte ich Märchen und Geschichten aus Europa. Im tiefen Inneren fühlte ich mich wie die Prinzessin im Urwald, die wandern muß, Beeren und Wurzeln essen, sich aus gefundenen Materialien bekleiden, bis sie wieder ihre Heimat findet. So entstand die Serie ›Hülsen‹, eine Reihe untragbarer Körper-Hüllungen.«

Eco-print describes a non-chemical dye system for textiles using plant material, which I am developing.

'By chance I was born in Australia. As a child I heard stories and fairy tales from Europe. Deep inside I felt like the princess wandering through a huge forest, eating roots and berries, and wearing clothes of found materials. This was the basis for the series ›husks‹, unwearable covers of the body.'

Hülse – Sommergarten
Wollfilz mit Flachs und Seide in
der Oberfläche. Eco-Print unter
Verwendung von verschiedenen
Rosensorten, Goldrute, Milch-
orange und Weide
L. 200 cm
1999
Husk – Summer Garden
Wool felt with flax and silk on sur-
face. Eco-Print using rosa species,
solidago, osage orange and willow
L. 200 cm
1999

Hülse – Winterreise
Wollfilz mit Seide, Baumwolle und
Flachs. Eco-Print unter Verwen-
dung von Hufnägeln, rostigen
Metallteilen und Pflanzensäften
L. 200 cm
1999
Husk – Winter Journey
Wool felt with silk, cotton and flax.
Eco-Print using horseshoe nails,
rusted metal fragments and plant
juices
L. 200 cm
1999

Hülse – Aus der Asche, Brot
Wollfilz, Eukalyptus-Farben
L. 200 cm
1999
Husk – Out of ashes, bread
Wool felt, eucalyptus dyes
L. 200 cm
1999

Jorie Johnson Kyoto (Japan)

Royal Chateau Brand
Wärmeset für Bewohner alter
Schlösser, speziell entworfen für
zugige Plätze
Ausführung für Prinzessinnen:
Nerz, Merinowolle
Ausführung für Prinzen: Silber-
fuchs, Merinowolle
Brille, Pappschachtel, Perlmutt-
knöpfe, Glasperlen, Stickerei
78 x 58 x 13 cm
1999
Royal Chateau Brand
A Cozy Kit for Inhabitants of Old
Palaces, specially created for drafty
environment
Princess Style: Mink fur and Merino
blend
Prince Style: Silver fox fur and
Merino blend
eye glasses, cardboard boxes, shell
buttons, beads, embroidery
78 x 58 x 13 cm
1999

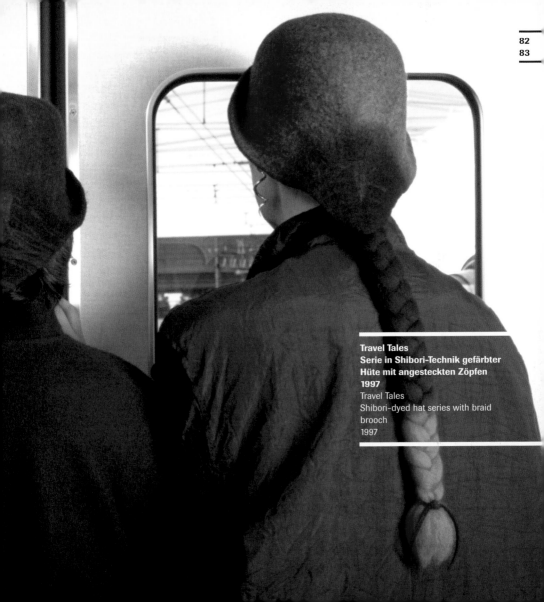

Travel Tales
Serie in Shibori-Technik gefärbter
Hüte mit angesteckten Zöpfen
1997
Travel Tales
Shibori-dyed hat series with braid
brooch
1997

Ingrid Thallinger Linz (A)

In den Raum greifen
Taschenobjekt
Schurwolle, gefilzt und gestrickt
80 x 50 cm
2000
Intervening in space
Pocket object
Pure new wool, felted and knit
80 x 50 cm
2000

**Ingrid Thallingers
Arbeit ist auf dem
schmalen Grat
zwischen funktio-
neller, tragbarer
Mode und künst-
lerischer Verrückt-
heit angesiedelt
und vereint beides.**
Ingrid Thallinger's
work walks the
tightrope between
functional, wearable
fashion and artistic
madness and unites
both qualities.
(Josef Aigner)

Behauptung
Kragen
Ärmlinge
Merinowolle, gefilzt, Spitzenstoff
2000
Headgear
Collar
Sleevelets
Merino wool, felted, lace material
2000

Filzen bedeutet, ganz aus dem Zentrum heraus zu arbeiten. Drei Zentimeter über dem Bauchnabel, aus dem Sonnengeflecht heraus.
Filzen heißt loslassen. Filzen kann nicht erklärt werden. Filzen geschieht. Je mehr Gespräche ich mit jeder dieser so verschiedenartigen Fasern geführt habe, desto besser können wir kommunizieren. Filzen ist urtümlich – ich muß mich ganz auf mich besinnen. Während des Filzprozesses entsteht die umhüllende Haut und wird sofort in ihre Dreidimensionalität geformt. Feinste Materialien kommen zum Einsatz. Merinowolle und der feine Spitzenstoff. Es entstehen durchlässige, leichte, umhüllende Filze.

Felting means working from the centre of one's being. Three centimetres above one's navel, from the solar plexus. Felting means letting go. Felting cannot be explained. Felting happens. The more talks I have had with each of these so different fibres, the better we can communicate. Felting is elemental – I have to take stock of myself wholly. The enveloping skin grows during the felting process and is shaped at once in its true three-dimensionality. The finest materials are used. Merino wool and fine lace material. Transparent, light, enveloping felts are created.

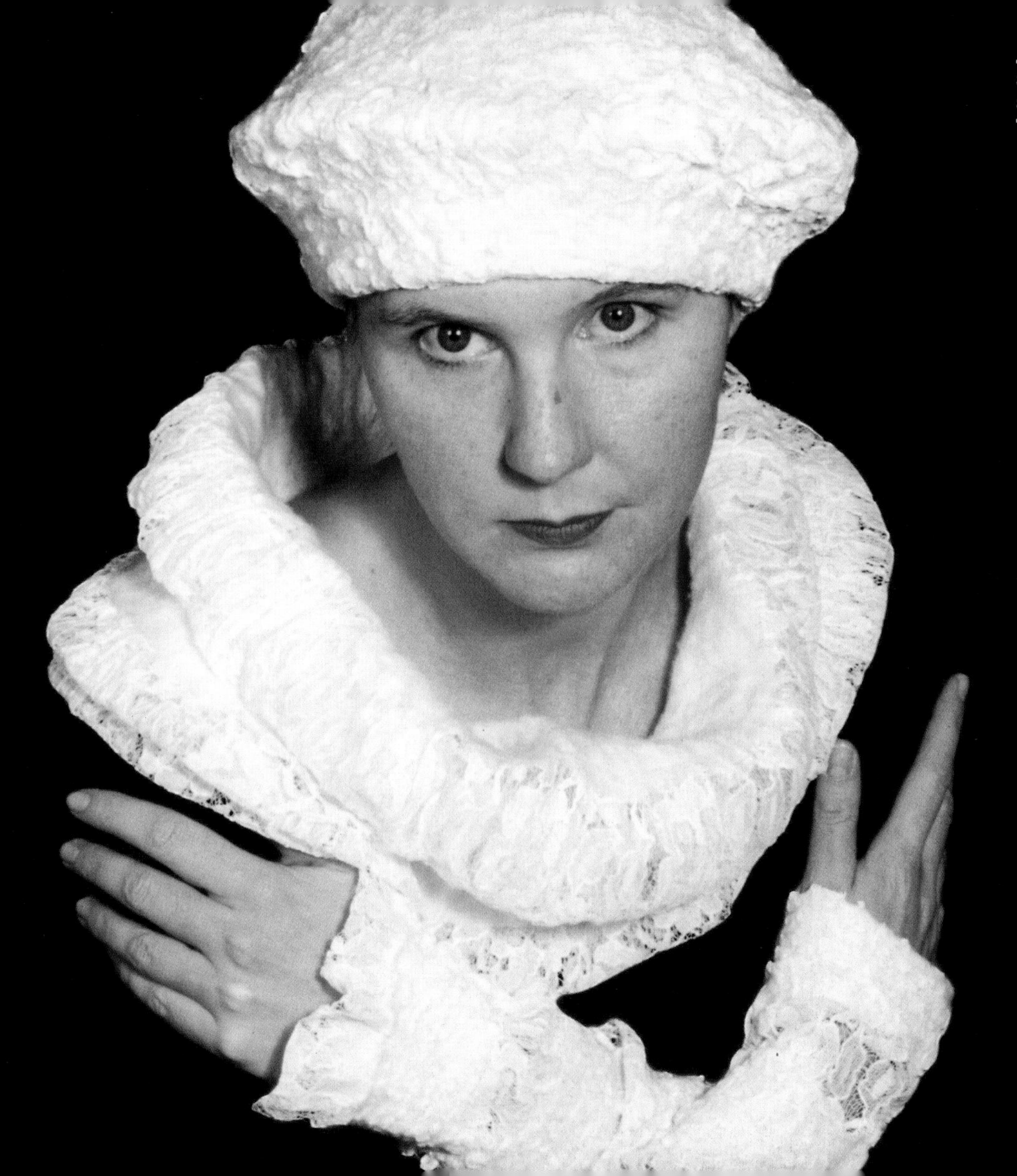

Hülle
Merinowolle, Seide, gefilzt
1997
Envelope
Merino wool, silk, felted
1997

Sweater
Merino-Wolle, Filz
75 x 150 cm
1998
Sweater
Merino wool, felt
75 x 150 cm
1998

Die Arbeiten von Heather Belcher haben stets einen Bezug zum Körper, wobei sie die Form abstrahierter Kleidungsstücke annehmen. Ihnen ist eine unfaßbare, traumartige Qualität eigen, und sie können als rätselhafte Figuren wie Torsi oder Fragmente von Gliedern gelesen werden. Die Textur des Stoffes steigert die Wahrnehmung der Kleidung als einer zweiten Haut und markiert eine Grenze zwischen dem Körper und dem Selbst.

Heather Belcher's work is concerned with ideas relating to the body, taking the form of abstracted clothing. They have a transient, dream-like quality and may be read as enigmatic characters, with torsos and limbs. The texture of the fabric itself heightens our awareness of clothing as a second skin, and marks a boundary between the body and one's self.

Célio Braga   Amsterdam (NL)

Ohne Titel II
Wolle, Seide, Tierhaare gefilzt;
Papiermaché, Gips, Ruß
190 x 16 x 30 cm
1999
Untitled II
Felt made from different kinds of
wool, silk, and other animal hair;
paper-maché, gesso, smoke
190 x 16 x 30 cm
1999

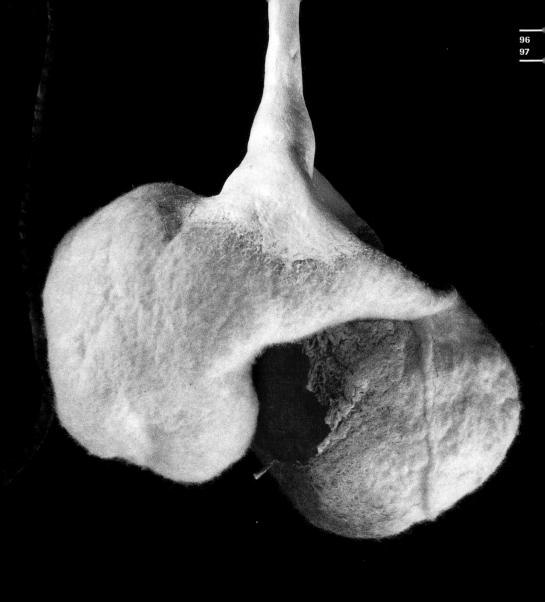

Die Fragmentierung des Körpers, das Zerlegen in Einzelteile, das Perforieren, das Verbinden seiner Teile auf neue Weise, das Verändern seiner Funktion und Symbolik: Célio Bragas Arbeiten beruhen auf der Darstellung des Körpers an einem Punkt, wo seine Organe, wo Haut und Blut Material für Versuchsanordnungen werden, die in die Gebiete der Kunst, der Wissenschaft, der Technologie und religiöser Verbildlichungen des menschlichen Körpers (Votivgaben) hineinreichen.

Die verwendeten Materialien sind meistens weich, einfach und geschmeidig und weisen dadurch deutliche Bezüge zu Haut und Fleisch auf. Die Arbeiten sind wie Öffnungen, Löcher, Narben oder Wunden geformt, die Behältnisse der Erinnerungen darstellen, in denen sich einmal erfahrene Verluste und Schmerzen lokalisieren; sie sind Bilder unserer Haut und Karten unserer Erfahrungen.

Das Wechselspiel zwischen hart und weich, innen und außen, die Verbindung der spezifischen Oberfläche von Filz mit anderen Materialien wie Samt, Papiermaché und Gips verleiht Bragas Arbeiten eine fremdartige und starke, an Eingeweide erinnernde Wirkung, die vom Schönen bis zum Grotesken, vom Sakralen zum Profanen reicht. Die Stärke von Bragas

Arbeiten liegt genau in der Widersprüchlichkeit des Wunsches, Schrecken und Schönheit gleichzeitig zu vollziehen und ihnen Gestalt verleihen zu wollen.

Fragmenting the body, splitting it up into pieces, perforating it, joining its parts in other ways, altering its function, its symbology: Célio Braga's works are based on the representation of the body where its organs, skin and blood become materials for experimentation, touching the issues of art, science, technology, and religious representation of the human body (votive offerings). The materials used for his works are mostly soft, humble and pliable materials which enhance its similarity to skin and flesh. The works are shaped accordingly to openings, holes, scars, and wounds which become containers of memories where loss and pain once took place, pictures of our skin and a map of our experiences.

The interplay between hard and soft, inside and outside combined with the sensibility of handmade felt and some other delicate materials, such as velvet, paper-maché, and plaster, gives the work a very strange and strong visceral quality that ranges from the beautiful to the grotesque, the sacred to the profane. The strength of Braga's pieces lays exactly in its contradictions and desire to fulfil and give shape both to horror and beauty.

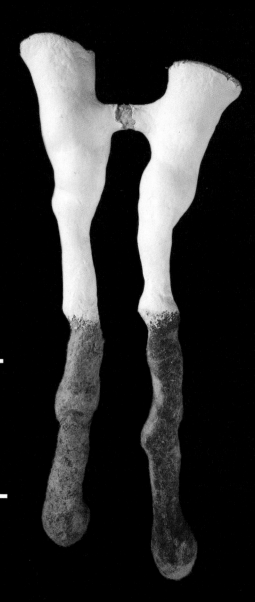

Ohne Titel III
Merinowolle, Seide gefilzt;
Papiermaché, Gips
40 x 13 cm
1999
Untitled III
Felt made form merino wool and
silk; paper-maché, gesso
40 x 13 cm
1999

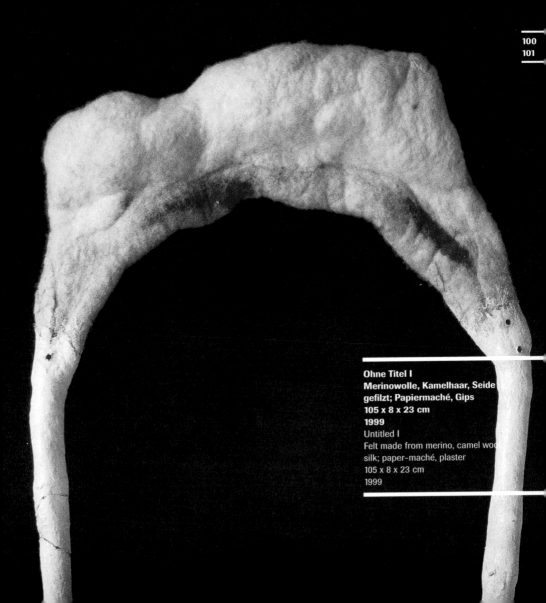

Ohne Titel I
Merinowolle, Kamelhaar, Seide
gefilzt; Papiermaché, Gips
105 x 8 x 23 cm
1999
Untitled I
Felt made from merino, camel wool
silk; paper-maché, plaster
105 x 8 x 23 cm
1999

Ingeborg Verres Karlsruhe (D)

Lichtfeld – Silent Landscape
Filzfliesen, geschichtet, Glasrund-
stäbe, Licht
Fläche: 200 x 200 cm,
Höhe: 180 cm
1997/98
Lichtfeld – Silent Landscape
Felt tiles, layered, round glass rods,
light
Surface: 200 x 200 cm,
Height: 180 cm
1997/98

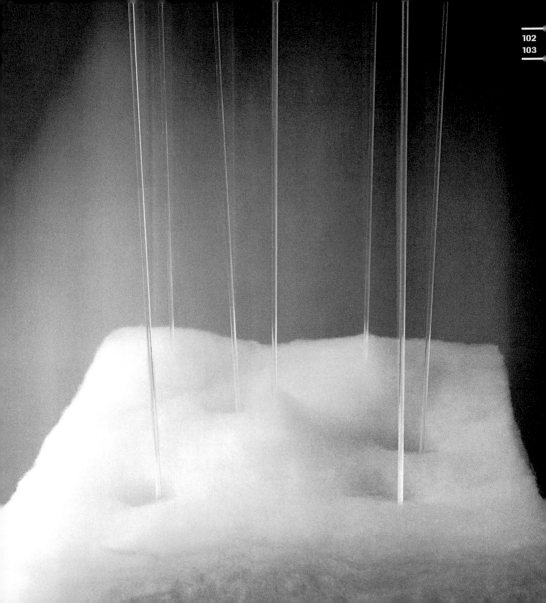

Das »Lichtfeld« besteht aus locker geschichteten weißen Wollfliesen, Glasstäbe leiten aus dem Inneren Licht an die Oberfläche. Da, wo sich die Wolle um einige der Glasstäbe herum einzieht, entstehen Mulden, wo sie sich an den Stäben hochwölbt, entstehen opake, lichtdurchwirkte Hügel. So wirkt die Oberfläche bewegt, sie pulsiert. Im oberen Bereich bleiben die Glasstäbe lichtlos. Nur an der Decke zeichnet sich durch Lichtkreise das innere Glühen ab.

The ›Light Field‹ consists in loosely layered white wool tiles; the light is lead out to the surface by glass rods. Where the wool puckers around some of the glass rods depressions are created; where it bulges up on the rods opaque hills interwoven with light are formed. Consequently the surface seems to move; it pulses. In the upper zone the glass rods remain lightless. Only on the ceiling is the inner glow revealed in circles of light.

»Silent Landscape« ist die Sprache von verschiedenartigen, nebeneinander bestehenden Dingen, die durch visuelle, musikalische Beziehungen zueinander ihre eigene Zeit und ihren eigenen, einmaligen Raum erschaffen.

In unserem Zeitalter der Elektrizität (Marshall McLuhan), der Gleichzeitigkeit und der Geschwindigkeit von Licht durchdringen sich Raum – Zeit – und Materie (Welt) gegenseitig. In dieser optimalen Wechselwirkung entstehen veränderte Formen von Energie, die zu technischen Erweiterungen und neuen Kommunikationsformen führen.

Die von Licht durchdrungene Materie, in der das Licht selbst die Information ist, schafft neue Formen der Wahrnehmung und verändert unsere Sinne. Die Gleichzeitigkeit, die Synchronizität, ermöglicht unmittelbares Erfassen. Sie erweitert die Einsicht in ursprüngliche organische Einheiten.

›Silent Landscape‹ is the language of different things that have been juxtaposed, which create their own time and their own, unique space by means of visual, musical relationships to each other.

In the age of electricity (Marshall McLuhan), synchronicity and the speed of light space – time – and material (cosmos) interpenetrate each other. In this best of all interrelationships transmuted forms of energy are created which lead to technical advances and new forms of communication. Material suffused with light, in which light itself is information, generates new forms of perception and changes our senses. Simultaneity, synchronicity, makes it possible to grasp something immediately. It enlarges insights into original organic unities.

Ulrike Hein Köln (D)

Model Series No. 01
Model Society
Filz
53 x 42 x 7 cm
1999/2000
Model Series No. 01
Model Society
Felt
53 x 42 x 7 cm
1999/2000

Verbotene Gegenstände
Filz, Stahlrahmen
L. 370 cm, B. 190 cm, H. 175 cm
1999
Forbidden Objects
Felt, steel frame
L. 370 cm, W. 190 cm, H. 175 cm
1999

Der Titel »Verbotene Gegenstände« basiert auf dem deutschen Waffengesetz § 37, in dem u. a. die Nachbildung und Bearbeitung von Kriegswaffen unter Strafe gestellt werden. Die Künstlerin Ulrike Hein bildet mit erforderlicher Genehmigung des Bundeskriminalamtes Wiesbaden (BKA) Kriegswaffen (Sturmgewehr AK 47-Kalaschnikow, Sturmgewehr HK-G3, Maschinenpistole UZI) originalgetreu in Filz ab. Filz steht hier für einen natürlichen Energiespeicher, der im Sinne der Künstlerin allgemein ein neutrales Energie- und Kraftfeld darstellt. In dieser (Objekt-)Installation werden potentiell tödliche Waffen durch Filz gleichförmig entwaffnet. Jede Ausstellung der nachgebildeten Kriegswaffen sowie auch ihr Verbleib müssen beim BKA angemeldet und von diesem ausdrücklich genehmigt werden: Verbotene Gegenstände – in Filz. (Joachim Rönneper)

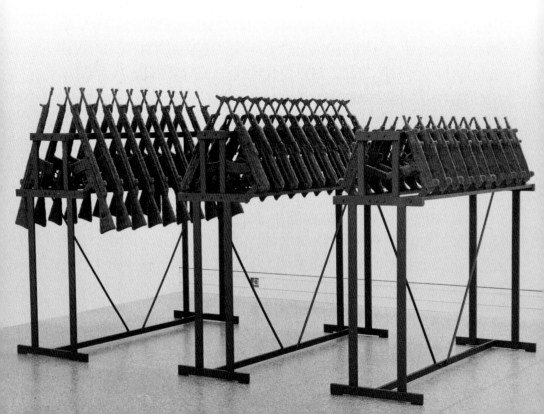

The title ›Forbidden Objects‹ is based on the German firearms law § 37, in which the reproduction and processing of military weapons is a punishable offence.

With the requisite permit from the Federal Office of Criminal Investigation Wiesbaden (BKA) the artist Ulrike Hein makes faithful reproductions of military weapons, such as the assault rifles AK 47 Kalashnikov and HK-G3 and the UZI pistol, in felt. Felt stands in this case for a natural energy accumulator which, as the artist views it, represents in general a neutral energy and charged field. In this (object) installation potentially lethal weapons are disarmed by being made in the same form but of felt. Every exhibition of these reproductions of military weapons as well as their whereabouts must be registered with the BKA and expressly approved: Forbidden Objects – in felt.

(Joachim Rönneper)

Ines Tartler Amsterdam (NL)

Filzschürze mit Unterhose
Filzschürze (Industriefilz) mit zwei
Ärmeln als Eingriff zum Körper,
weiße Feinrippunterhose
Schürze: ca. 80 x 35 cm
Unterhose: 36 x 31 cm
1999
Felt apron with underpants
Felt apron (industrial felt) with two
sleeves as an intervention in the
body, white finely ribbed underpants
Apron: ca. 80 x 35 cm
Underpants: 36 x 31 cm
1999

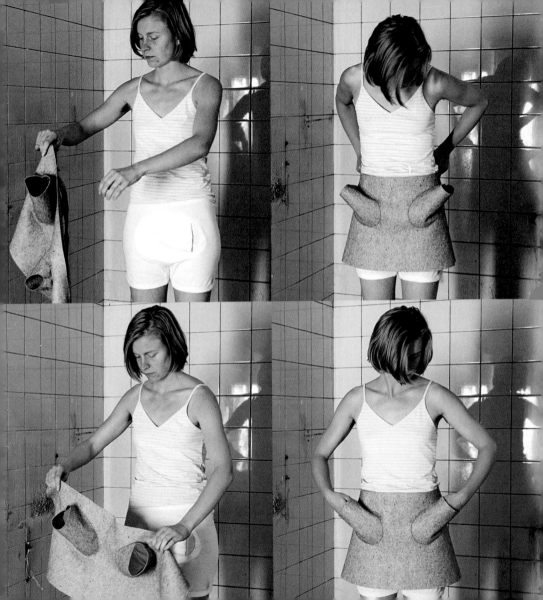

**Verdichtetes, komprimiertes, organisches Material.
Ein Material als metaphorisches Werkzeug. Entstanden
durch eine Handlung, die einen Zeitraum und einen
Prozeß abzwingt. Das Resultat ist zugleich Momentauf-
nahme und Handlungskonsequenz, mit der bleibenden
Möglichkeit der weiteren Formung. Wie ein Aggregat-
zustand verbleibt dieser taktile Wärmeisolator in seiner
zeitlichen Erscheinung. Ist jede Einstülpung auch eine
Ausstülpung. Die Position eine Konsequenz. Angehal-
tene Momente, sinnbildlich für die Möglichkeit von Veränderung. Eine Handreichung, eine Unterstreichung
einer unbemerkten täglichen Handlung, eine introver-
tierte, stille Handlung – nicht mehr und nicht weniger.
Wie eine Frage bleiben diese Hüllen zurück.**

Condensed, compressed organic material. A material as a
metaphorical tool, created by an act which compels a space
of time and a process. The result is at once a snapshot and
a consistent plot with the possibility of further shaping
remaining. Like a state of matter, this tactile warmth insulator
remains in its temporal appearance. Each turning in is also
turning out. Position a consequence. Arrested moments, sym-
bolising the possibility of change. Giving a hand, underscor-
ing an unnoticed daily act, an introverted, quiet act – no
more, no less.
These envelops remain behind like a question.

(Iris Eichenberg)

Belichtungszeiten
Filz, lichtgebleicht
300 x 310 cm
1996
Exposure times
Felt, faded by light
300 x 310 cm
1996

Den Ausgangspunkt der Arbeit bilden die originalen Wandbespannungen aus der Gemäldegalerie Wilhelmshöhe in Kassel. Auf dem roten Filz haben die Gemälde des Museums wie auch die Schilder aus Plexiglas Spuren hinterlassen. Der Filz um die Bilder herum ist in mehr als 20 Jahren durch die starke Lichteinwirkung ausgeblichen. Zu sehen sind nunmehr die tiefroten Schatten der Bilder, teilweise sind die Schilder noch lesbar – je nachdem, ob die Wände nach Süden oder Norden standen. Ebenfalls sichtbar sind schwarze Spuren, die von der Metallkonstruktion der Wand herrühren, die Staub anzogen, manchmal sieht man auch Spuren von Wasserschäden.

Anfänglich machte ich Fotoaufnahmen von diesen außergewöhnlichen Schatten, die zu sehen waren, wenn ein Gemälde z. B. zur Restaurierung abgehängt war. Später (Ende 1995), als das Museum umgebaut werden sollte, konnte ich die ersten originalen Filze erhalten.

Diese Filze sind von mir auf Rahmen aufgespannt worden. Sie stehen auf dem Boden, um die Ambivalenz von Wand und Bild zu betonen.

The original wall hangings from the Wilhelmshöhe Paintings Gallery in Kassel are the point of departure for this work. The paintings as well as the perspex labels in that museum left traces on the red felt there. The felt round the pictures faded over more than 20 years from the effects of bright light. Now what is visible are the dark red shadows of the pictures; some of the labels are still legible – depending on whether the walls faced south or north. What is also visible are the black traces which were left by the metal construction of the walls because they attracted dust; sometimes traces of damage caused by water are discernible.

In the beginning I took photos of these extraordinary shadows which became visible when a painting was being restored, for instance. Later (at the close of 1995), when the museum was scheduled to be remodelled, I was able to obtain the first original pieces of felt. I stretched these pieces of felt on stretchers. They stand on the floor to emphasize the ambivalence of wall and picture.

Verena Gloor   Zürich (CH)

Kreuzteppich – Ort der Begegnung
Dunkelbraune und wenig dunkelblau
eingefärbte Walliser Wolle,
zusammengekardet.
205 x 205 cm pro Kreuz,
insgesamt ca. 400 x 400 cm
1990
Mitarbeit: Margreth Edlin, Kathrin
Gusset und Claudia Hunger
Cross carpet – Meeting place
Dark brown and some dark blue dyed
Valaisan wools, carded together.
205 x 205 cm per cross,
overall approx. 400 x 400 cm
1990
In collaboration with: Margreth Edlin,
Kathrin Gusset and Claudia Hunger

Ich hatte mit Freunden eine romanische Kirche in Graubünden besucht. Angeregt durch dieses gemeinsame Erlebnis, sah ich unvermittelt die Idee eines Teppichs: ein Kreuz.

Das gleichschenklige Kreuz ist ein uraltes astronomisches Zeichen, u.a. für die Kennzeichnung der vier Himmelsrichtungen und den Lauf der Sonne durch die vier Jahreszeiten. Es ist ein echtes, altes Wandlungs- und Lebenssymbol, ein Symbol für Rhythmus und Maß.

I had visited a Romanesque church in the Grisons with friends. Inspired by this communal experience, I immediately envisaged the idea for a carpet: a cross.

The cross with arms of equal length is an ancient astronomical sign denoting the four points of the compass and the course taken by the Sun through the four seasons. It is a genuine, old symbol of change and life, a symbol of rhythm and measure.

Ohne Titel
Achtteilige Bodenarbeit, variabel
kombinierbar
Industriefilz, geklebt, gesägt,
geschoren
ca. 35 x 120 x 120 cm
1999
Untitled
Eight-piece floor object, combinable
as desired
Industrial felt, glued, sawn, shorn
approx. 35 x 120 x 120 cm
1999

Filz & Form

Form und Material sind wesentliche Bedeutungsträger im Werk von Ernst Köster. Sie stehen immer in einem Spannungsverhältnis, das aus ihren strukturellen Eigenschaften erwächst. Bestimmte Formen scheinen unabdingbar mit einem Material verbunden, wiederum evozieren bestimmte Materialien formale Notwendigkeiten. Anstelle des Bleistifts benutzt Köster den Fotoapparat zum Festhalten von Ideen, sein Fotoarchiv ersetzt das Skizzenbuch. Formmotive werden gesammelt, gesichtet – aus der Gestaltungsvielfalt kristallisieren sich Grundformen und künstlerische Konzepte heraus.

Thema des Künstlers ist es, aus der alltäglichen visuellen Fülle formale Konstanten mit elementaren Qualitäten aufzuspüren und herauszufiltern. Ihn interessiert, Formen zu entdecken und nicht zu erfinden. Ernst Köster versucht »Formen zu finden, die es ganz häufig gibt, die aber in den verschiedensten Zusammenhängen erscheinen«. So entstehen Objekte, Skulpturen oder Bilder, deren Motive und Materialien von vielfältigsten Zusammenhängen geprägt sind, die aber immer auf die formalen Grundelemente unserer visuellen Kommunikation zielen.

Kreuz und Zylinder, diese plastischen Volumen sind nicht allein als konkrete Form zu verstehen. Gepaart mit den Materialeigenschaften von Filz entstehen hier Objekte, die durch ihre klare und einfache Erscheinung mehrdeutig werden und ein weites semantisches Feld eröffnen. Die sinnliche Qualität der Filzobjekte wird durch den Kontrast von Material und Form gesteigert. Dem warmen, weichen Filz mit chaotischer Faserstruktur werden scharf konturierte geometrische und regelmäßig geordnete Formen zugesellt, die vielfältige Assoziationen auslösen und den Betrachter zur individuellen Kontextsuche animieren.

Die serielle Bearbeitung der Kreuzform als Positiv- wie Negativform und das variable Kompositionsprinzip verdeutlichen spielerisch die unerschöpflichen Möglichkeiten, die durch Variation, Modifikation & Kombination im elementaren Formenrepertoire enthalten sind.

Köster reduziert komplexe Wahrnehmungsstrukturen zu autonomen, sinnlich ansprechenden Bildzeichen, die ihren Kontext sowohl in der visuellen Ästhetik des Alltäglichen wie der Kunstgeschichte finden können.

(Ariane Hackstein)

Felt & Form

Form and material are essentially loaded with meaning in Ernst Köster's work. They are always in a relationship to one another that is fraught with tensions, growing out of their structural properties. Certain forms seem to be linked irrevocably with a material; then again other materials evoke formal exigencies. Köster uses a camera instead of a pencil to capture ideas; his photo archive replaces the sketchbook. Form motifs are collected, viewed – basic forms and artistic concepts are the precipitate of a diversity of form.

The artist's theme is tracking down formal constants with elemental properties from the wealth of visual impressions encountered every day and filtering them out. He is interested in discovering, not in inventing, forms. Ernst Köster is trying to ›find forms which occur frequently but appear in the most diverse contexts‹. Thus objects, sculptures or pictures come into being with motifs and materials shaped by the most diverse contexts but aiming invariably at the basic formal elements of visual communication with one another.

The cross and cylinder, plastic volumes like these are not to be viewed solely as a concrete form. Coupled with the material properties of felt, objects emerge here which are ambivalent in meaning and open up a wide semantic field. The sensual quality of felt objects is enhanced by the contrast between material and form. Sharply contoured, regularly ordered geometric shapes are made to accompany warm, soft felt with its chaotic fibre structure, which trigger off diverse associations, inspiring the viewer to search for individual contexts.

Serially working on the cross shape as both a positive and a negative form and the variable principle of composition are playful demonstrations of the inexhaustible possibilities available in the elemental repertoire of forms through variation, modification and combination.

Köster reduces complex perceptual structures to autonomous, sensuously appealing pictorial signs which can find their context both in the visual aesthetic of the everyday and in art history.

(Ariane Hackstein)

Hands
Filztafelbild
77 x 89 cm
1999
Hands
Felt panel picture
77 x 89 cm
1999

Circulation
Filztafelbild
93 x 100 cm
1999
Circulation
Felt panel picture
93 x 100 cm
1999

Man Sees Tree
Filztafelbild
88 x 105 cm
1999
Man Sees Tree
Felt panel picture
88 x 105 cm
1999

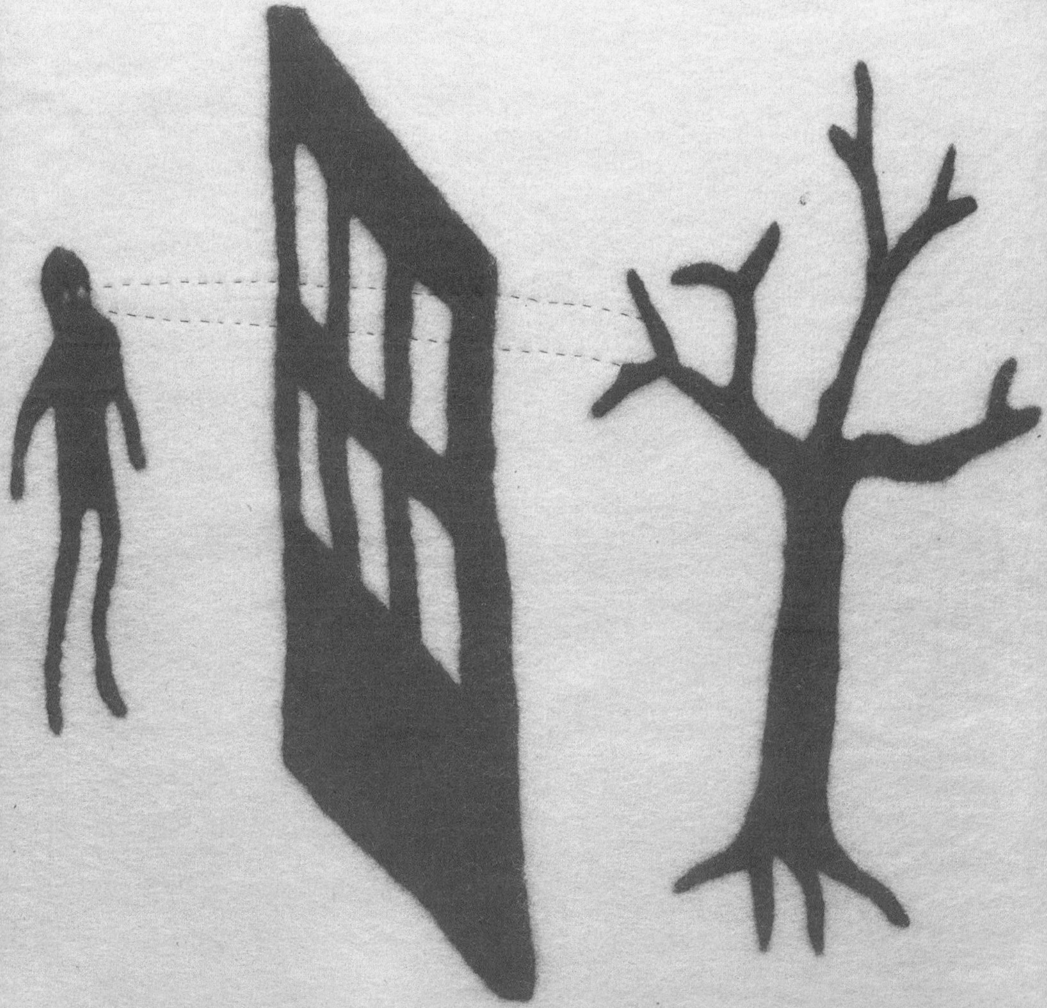

»Flexi«
Schattenfigur
Filz
230 x 125 cm
1999
»Flexi«
Shadow figure
Felt
230 x 125 cm
1999

»Figure«
Filz, Gewebe mit Silber
200 x 53 cm
1999
»Figure«
Felt and material with silver
200 x 53 cm
1999

»Flexi« ist ein ganz einfaches Werk, bei dem jedoch die richtige Beleuchtung entscheidend ist: Die langen Filzröhren sind mit einem Abstand an der Wand befestigt, so daß sie Schatten werfen können. Die Idee ist, die Arbeit jeweils an Ort und Stelle nach den vorgegebenen Maßen der Wand zu komponieren. Die Werkstücke heißen »Flexi«, weil sich ihre Formation niemals genau wiederholt.

»Flexi« is a very simple work, instead the correct lighting becomes important. The work is composed of long felted tubes. It is put up on the wall but with a distance to throw shadows. It is the idea to compose the work on the spot or according to given measures of the wall etc. It is called »Flexi« because it will never be repeated in exactly the same way again.

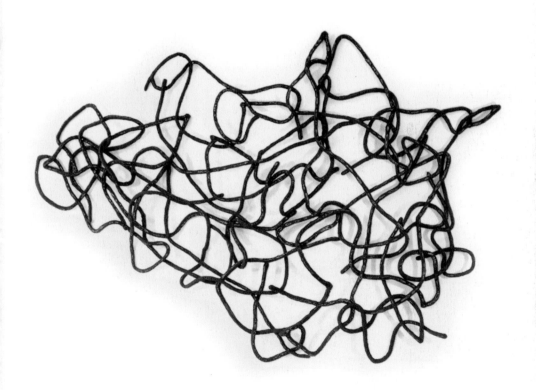

Schweres und Leichtes
Wollfilz
H. 210 cm, B. 140 cm, T. 170 cm
1999/2000
Heavy and light
Wool, felted
H. 210 cm, L. 140 cm, W. 170 cm
1999/2000

Vorbild für meine Arbeit sind mittelalterliche Grabmäler mit der steinernen Abbildung des Verstorbenen als Grabplatte.
Erde – Luft, Körper – Seele, Erdenschwere – Himmelsleichtigkeit: zwei verschiedene Filzarten unterstützen die Aussage: einmal die Wolle dick, wulstig, unregelmäßig gefilzt, als Platten aufeinandergehäuft: an Stein erinnernd, aber doch fast noch Kleidungsstück – das Textile als Bindeglied zwischen Leben und Tod – auch tröstlich: warm und weich statt kalt und hart. Und dann die Wolle hauchdünn, transparent gefilzt: es ergeben sich Assoziationen zu Luftigem, Fliegendem, Spinnweb – zarteste Form von Materie.

Medieval tombstones with the stone portrait of the dead as a tomb slab are the models for my work.
Earth – air, body – soul, earthy heaviness – heavenly lightness: two different types of felt underscore the statement: first the wool, thick, bulging, irregularly felted, as slabs piled on top of each other: recalling stone but still almost a piece of clothing – the textile quality as the link between life and death – also consoling: warm and soft instead of cold and hard. And then the wool, felted whisper-thin, transparently: associations with what is airy, flying, spider web arise – the most delicate form of material.

Mond
Tasche, in einem Stück gefilzt
Merinowolle, gefärbt, Mohair,
gefärbt
Ø ca. 70 cm
1999/2000
Hergestellt in Zusammenarbeit
mit Maria Mahlmann
Moon
Bag, felted in one piece
Merino wool, dyed; mohair, dyed
Ø ca. 70 cm
1999/2000
Made in collaboration with Maria
Mahlmann

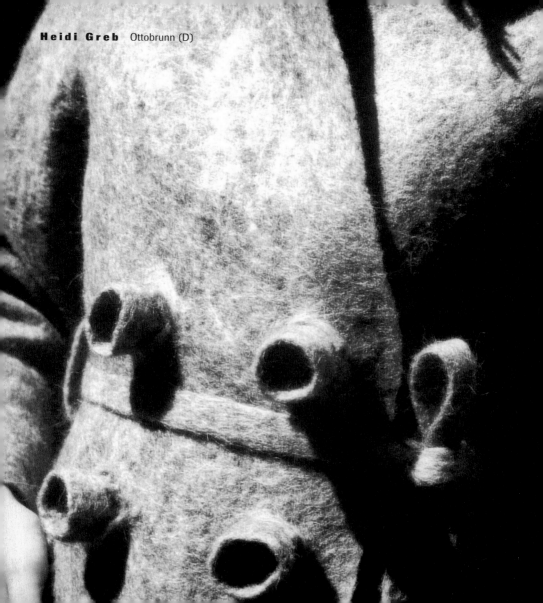

Heidi Greb Ottobrunn (D)

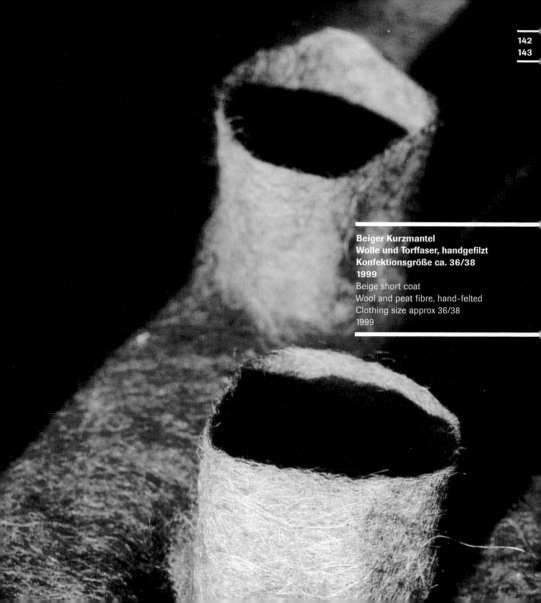

Beiger Kurzmantel
Wolle und Torffaser, handgefilzt
Konfektionsgröße ca. 36/38
1999
Beige short coat
Wool and peat fibre, hand-felted
Clothing size approx 36/38
1999

Gunilla Paetau Sjöberg Bålsta (S)

Flake of Soot
Wolle, Seide, Flachs, Baumwolle
56 x 260 cm
1999
Flake of Soot
Wool, silk, flax, cotton
56 x 260 cm
1999

Flake of Sorrow
Wolle, Seide, Flachs, Baumwolle
56 x 260 cm
1999
Flake of Sorrow
Wool, silk, flax, cotton
56 x 260 cm
1999

Der Wolle als Erwachsene zu begegnen, bedeutete eine Reise zurück in meine Kindheit und ihre sinnlichen Erfahrungen und zugleich den Beginn einer Erkundungsfahrt in die künstlerischen Möglichkeiten des Materials Wolle und der alten Technik des Filzens. Zunächst fand ich, daß plastische Formen zu erforschen ein faszinierendes Thema war. Der Wolle drei Dimensionen zu verleihen: Volumen und Körper! Ich stellte Versuche an, die Sinnlichkeit des Materials, die lebendige Ausdruckskraft seiner Haptik zu bewahren, während ich mich gleichzeitig bemühte, den Objekten eine klare Form zu geben.

Nachdem ich mich viele Jahre mit volumenreichen Plastiken beschäftigt hatte, habe ich mich nun dünnen, zweidimensionalen Arbeiten zugewandt. Das Zarte, Fragile, Transparente übt auf mich eine besondere Anziehung aus.

›Rußflocke‹ und ›Sorgenflocke‹ sind Teil eines Projekts mit dem Titel ›Feuerstelle‹. Wir alle entflammen, brennen, in Liebe, in Leidenschaft, in Freundschaft. Das Projekt erzählt vom Leben einer Frau. Es erzählt von existenziellen Erfahrungen, von Leben und Tod.

Seit einiger Zeit beschäftige ich mich auch mit Raumkonzepten – es interessiert mich, mit meinen Arbeiten einen Raum zu definieren. In einem Raum können die Arbeiten zueinander in eine Beziehung treten, zwischen ihnen findet ein Prozeß statt. Die Arbeiten sind wie einzelne Gedanken; sobald sie einen Raum finden, ist es, als bekämen sie einen Rahmen, in dem sie wachsen können. In einem Raum beieinanderstehend oder -hängend können sie mir von geheimen Verbindungen und Zusammenhängen erzählen, die ich vorher nicht bemerkt habe. Und der Raum kann von einem besonderen Bild, einem Gefühl, einer Botschaft erfüllt sein.

Meeting wool as an adult turned into a trip back to my childhood and its sensual experiences, and an exploration into the artistic possibilities of the material wool and the old technique of felt-making. At first I found that sculptural forms were a very fascinating issue to explore. To give the wool three dimensions: volume and body! I also experimented to preserve the material's sensuality, its living and tactile expression, in the same way trying to give the object a distinct form.

After having worked with thick sculptures for many years I have turned to thin and two dimensional works. There is a special attraction in thinness, fragility, and transparency.

›Flake of Soot‹ and ›Flake of Sorrow‹ are part of a concept called ›Place of Fire‹. We all inflame, burn in love, in passions, in friendship. It is a story of a woman's life. It is a story of essential matters, life and death.

My interest in the room as a concept is also growing – to formulate the room with my works. In a room the works can find a relation to each other, there is a process going on between them. The works are like separate thoughts; whenever they find a room it is as when they got a body to grow in. Hanging/standing together in the room they can tell me their secrets, connections and contexts I had not seen before. And the room can be filled with a special image, feeling, message.

Das Maul
**Speziell entwickelte Edelstahlfilz-
technik in Kombination mit verschiedenen Kaltverformungs- und
Härteprozessen.**
H. 19 cm, B. 45 cm, T. 25 cm
1999
The Mouth
Specially developed technique for
making stainless steel felt combined
with various cold-working and tempering processes
H. 19 cm, W. 45 cm, D. 25 cm
1999

Ich schaffe Gefäßobjekte aus Edelstahlfilz. Die Anwendung der traditionellen Filztechnik mit dem zeitgenössischen Material Edelstahl ergibt in Verbindung mit speziellen Härte- und Schmiedeprozessen dreidimensionale, massive Edelstahlfilzkörper.

Das harte, widerspenstige Material Edelstahl verbindet sich hier mit der weichen, haptischen Ausstrahlung des Filzes. Prozesse zur Veränderung der Farbigkeit ermöglichen »farbräumliche« Akzentuierungen.

Meine Gefäßobjekte sind Behältnis für etwas, was nicht hortbar ist, für Licht, Luft, Stimmungen und Gefühle.

I make vessel objects of stainless steel felt. Using a traditional felting technique on the contemporary material stainless steel results, in combination with special tempering and forging processes, in three-dimensional, solid stainless steel felt objects.

The hard, intractable material that is stainless steel is here associated with the soft, tactile aura of felt. Processes for changing its colour make ›colour-spatial‹ accentuation possible.

My vessel objects are receptacles for something which can not be hoarded, for light, air, moods and feelings.

Sea Maid
Weste, Bluse Hut, Leggings
Mit Anilinfarben in einem
Zwischenstadium des Filz-
prozesses von Hand gefärbt.
Handgefilzt, ohne Nähte
2000
Sea Maid
West, skirt, hat, leggings
Hand dyed with chemical dyes on
prefelt stage in the felting prosess.
Handfelted without stitches
2000

Triumph der Sonne
Merinowolle, verschiedene Garne
und Gewebe
87 x 44 cm
1999
The Sun's Triumph
Merino wool, different yarns and
fabrics
87 x 44 cm
1999

Inge Evers   Haarlem (NL)

Memento
Australische Merinowolle, Viskose
bedruckt, Seide, Knöpfe, Leinen-
stickerei, Baumwollfutter
90 x 160 cm
1999
Mit freundlicher Genehmigung
Mary E. Burkett
Memento
Australian Fine Merino Wool, viscose
printed, silk, buttons, linen stitches,
cotton lining.
90 x 160 cm
1999
Courtesy Mary E. Burkett

Mammas (Detail)
Australische Merinowolle, Viskose
bedruckt und gefärbt, Leinen-
stickerei, Baumwollfutter
90 x 160 cm
1999
Mammas (Detail)
Australian Fine Merino Wool, viskose
printed and tie dyed, linen stitches,
cotton lining
90 x 160 cm
1999

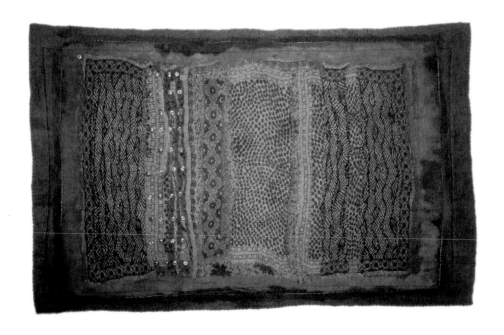

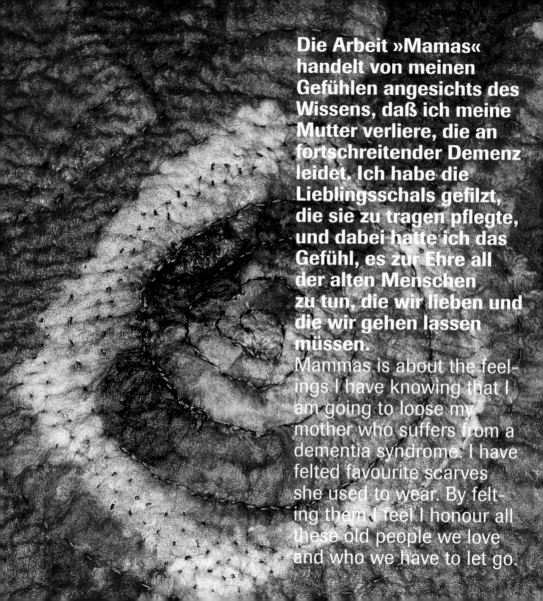

Die Arbeit »Mamas« handelt von meinen Gefühlen angesichts des Wissens, daß ich meine Mutter verliere, die an fortschreitender Demenz leidet. Ich habe die Lieblingsschals gefilzt, die sie zu tragen pflegte, und dabei hatte ich das Gefühl, es zur Ehre all der alten Menschen zu tun, die wir lieben und die wir gehen lassen müssen.

Mammas is about the feelings I have knowing that I am going to loose my mother who suffers from a dementia syndrome. I have felted favourite scarves she used to wear. By felting them I feel I honour all these old people we love and who we have to let go.

Anlaß für die Arbeit
»Momento« war ein Hub-
schrauberflug durch eine
gebirgige Landschaft in
Georgien. Gleich hinter
dem nächsten Gipfel fand
der Tschetschenienkrieg
statt. Die Herzen der
Leute waren warm wie
die Sonne, es gab eine
schmerzliche, struktu-
relle Harmonie zwischen
Krieg und Frieden.

Memento is about flying
a helicopter through a
mountainous landscape in
Georgia. The Tsjetsjen war
is just over the next top.
The hearts of the people are
warm as the sun, there is a
painful and structured har-
mony between war and
peace.

Claudia Merx Stolberg (D)

Zur Ruhe gekommen
Raumfilz, zweiseitig
Wolle, handgefilzt
140 x 195 cm
1998
Gone to rest
Spatial felt, two-sided
Wool, hand-felted
140 x 195 cm
1998

Die beidseitigen dünnen Raumfilze sind ge-
prägt durch die Beziehung der Muster beider
Seiten, die sich durchdringen und in Struktur
und Farbigkeit gegenseitig beeinflussen.
The two-sided, thin spatial felts are distinguished
by the relationship of the patterns of both sides
which interpenetrate each other, influencing one
another in respect of structure and colour.

Lene Frantzen Kolding (DK)

Memories of a Pencil
Wolle gefilzt
190 x 380 cm
1999
Memories of a Pencil
Felted wool
190 x 380 cm
1999

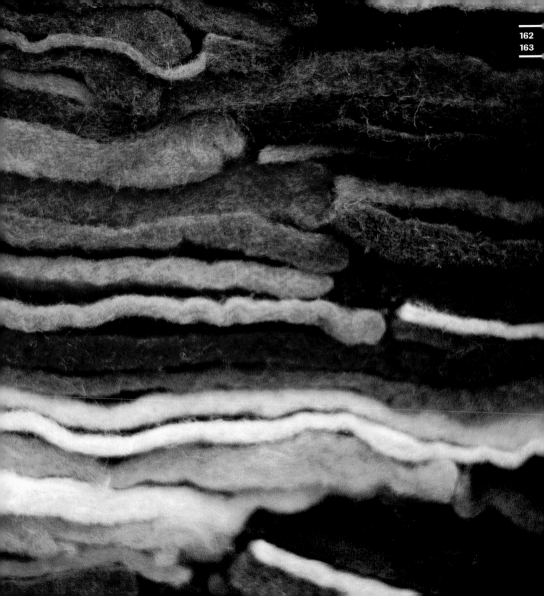

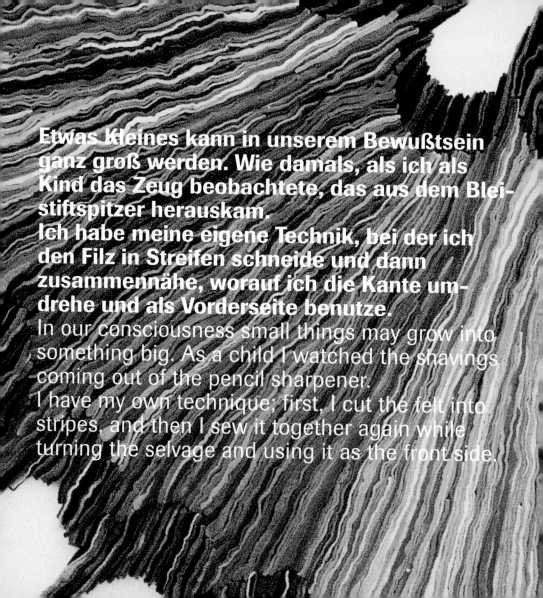

Etwas Kleines kann in unserem Bewußtsein ganz groß werden. Wie damals, als ich als Kind das Zeug beobachtete, das aus dem Bleistiftspitzer herauskam.
Ich habe meine eigene Technik, bei der ich den Filz in Streifen schneide und dann zusammennähe, worauf ich die Kante umdrehe und als Vorderseite benutze.

In our consciousness small things may grow into something big. As a child I watched the shavings coming out of the pencil sharpener.
I have my own technique; first, I cut the felt into stripes, and then I sew it together again while turning the selvage and using it as the front side.

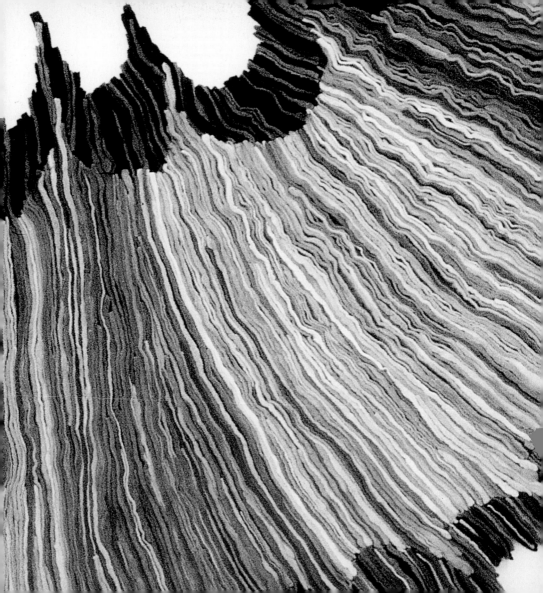

Käthi Hoppler-Dinkel   Ostermundigen (CH)

Melusinenschuhe
Weiße Merinowolle, verschiedene
Kardwolle, gefärbt, Wollfransen
1990
Melusinenschuhe
White merino wool, different card
wool, dyed, wool-fringes
1990

Schritt-für-Schritt-Stiefel
Schwarze Walliser Wolle, Merino-
wolle, gefilzt
70 x 25 x 50 cm
2000
Schritt-für-Schritt-Stiefel
Black Valaisan wool, merino wool,
felted
70 x 25 x 50 cm
2000

Eiliger Wächterstiefel
Verschiedene Kardwolle, gefärbt
70 x 30 x 50 cm
1995
Eiliger Wächterstiefel
Different card wool, dyed
70 x 30 x 50 cm
1995

Ich bewege mich mit meinen Arbeiten immer innerhalb des Themas
›Bekleidung‹. Filz muß für mich ›hautnah‹ sein.
Während mehrerer Jahre setzte ich mich mit größeren Gegenständen
(z. B. Jacken) auseinander, konzentrierte mich dann aber immer mehr
auf ›Fuß-Hüllen‹. Seit 1987 produziere ich ausschließlich Filzschuhe.
Diese sind immer ›aus einem Guß‹ hergestellt, d.h. ohne Nähte.
Meine Filzschuhe sind als Muße-Schuhe gedacht. Sie können als
Objekte betrachtet oder als wärmende, sich dem Fuß anschmiegende
Hülle getragen werden.
Nicht alle meine Schuhe können fliegen. Es kommt auch vor, daß sie
ganz allmählich, einen Schritt vor den anderen setzend, in ihre Gestalt
hineinwachsen. Schließlich stehen sie da, zu jedem Schritt verlockt, für
jeden Schritt belohnt, keinen Schritt missend.

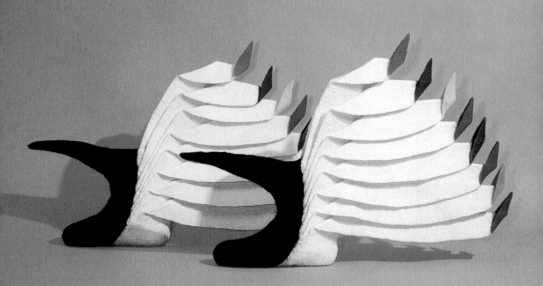

With my work I'm always moving within the theme of ›clothing‹. For me felt must be »as close as skin«.

For several years I was preoccupied with largish objects (e.g. jackets), but then increasingly with »foot-envelopes«. Since 1987 I've been producing nothing but felt shoes. These are always »from one mould, a perfect whole«, that is, no seams.

My felt shoes are conceived as leisure shoes. They can be viewed as objects or worn as an envelope that nestles on the foot.

Not all my shoes can fly. It can also happen that they grow into their form only gradually, taking one step at a time. Ultimately they stand there, enticed to each step, rewarded for each step, not missing a step.

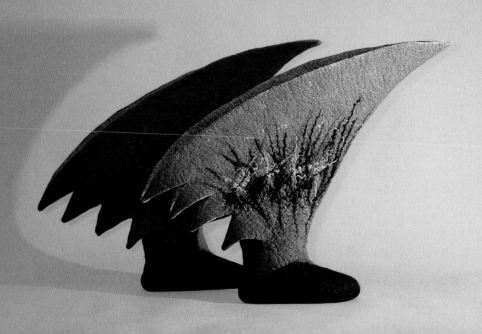

Riona Kuthe Nürtingen (D)

Nr. 15.99
Filz
150 x 150 x 14 cm
1999
Nr. 15.99
Felt
150 x 150 x 14 cm
1999

Masken verhüllen, verheimlichen, täuschen. Sie kommen auf der ganzen Welt als frühe Zeugnisse der Kulturgeschichte vor. Ihr Sinn und ihre Funktion sind so vielfältig wie ihre Form und Gestaltung. Der, der sie trägt, wird nicht erkannt und ist geschützt. Damit könnte man Filz und Masken als eine ideale Verbindung ansehen, da auch der Filz seit Urzeiten den Menschen als Schutz dient.

Masks disguise, conceal, deceive. They occur throughout the world as early documents of cultural history. Their meanings and their functions are as diverse as their shapes and designs. The person who wears one is not recognized and is, therefore, protected. This means that felt and masks can be viewed as an ideal match since felt, too, has protected humankind from time immemorial.

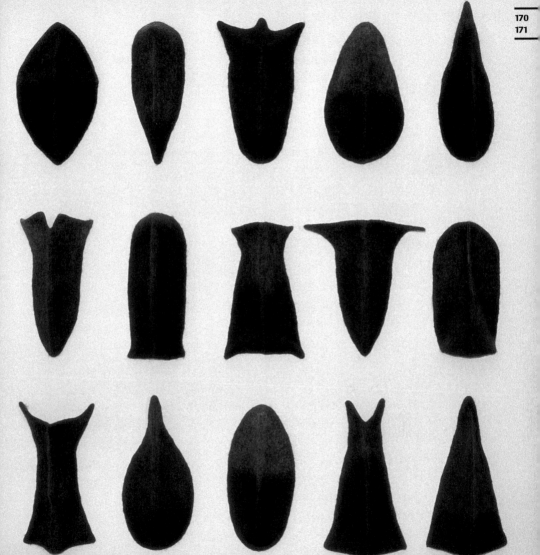

Küchenzauber oder Das Haus der Baba Yaga
12 Filzobjekte, die in unterschiedlicher Zahl und Anordnung auf einem Küchentisch arrangiert werden können.
Handgefertigter Wollfilz; Dekorationen u.a. Stoffe, Perlen, Rosshaar.
Die Standvorrichtungen der Objekte bestehen aus diversen Haushaltsgegenständen.
Tischmaße: H. 79 cm, B. 97 cm, T. 62 cm; Höhe der Objekte: 11–54 cm
1999
Kitchen magic or Baba Yaga's House
12 felt objects, various of which can be rearranged on a kitchen table
Hand-made wool felt; decorations include materials, pearls, horsehair.
The stands for the objects consist in various household objects.
Table dimensions: H. 79 cm, W. 97 cm, B. 62 cm; Height of objects 11–54 cm
1999

Die meisten Handfilz-Hersteller sind Frauen, ihre »Werkstatt« meist der Küchentisch, und nicht selten werden ihre Werke als »Hausfrauenkunst« abgetan. Küchentisch, diverse Haushaltsutensilien dieser Installation nehmen darauf Bezug. Im Kontext dieser »Küchen-Werkstatt-Inszenierung« wird den Utensilien nun eine neue Funktion zugewiesen, ein ungewöhnlicher Spielraum eröffnet. Auch die Objekte, für die sie als Träger fungieren, greifen dieses Thema der Wandelbarkeit auf. Als Vorbild der Figuren dient das Haus der Baba Yaga, einer Märchenfigur aus dem osteuropäischen Raum, unserer Hexe vergleichbar, das auf Hühnerbeinen tanzt und nicht auf dem Boden, nicht einem fixen Standpunkt verhaftet ist.
Diese Beweglichkeit habe ich als Spielraum aufgegriffen. Meine Arbeit ist variabel – viele Blickwinkel und Interpretationen sind möglich. Vielleicht, daß sie Anstoß gibt, über Begriffe wie »Hausfrauenkunst« nachzudenken.

Most makers of felt by hand are women; their ›workshops‹ are usually kitchen tables and quite often their work is derided as ›housewives' art‹. The kitchen table and various household utensils in this installation allude to this. In the context of this ›kitchen-workshop production‹ the utensils are assigned new functions; unusual scope is opened up.

The objects for which they function as supports also take up the theme of mutability. The house of Baba Yaga, an Eastern European fairy-tale figure comparable to our witches, is the model for these figures. Her house dances on chicken legs and does not rest on the ground, is not anchored to any fixed point.

I have taken advantage of this mobility as affording scope. My work is variable – many perspectives and interpretations are possible. Perhaps it may stimulate people to think about terms like ›housewives' art‹.

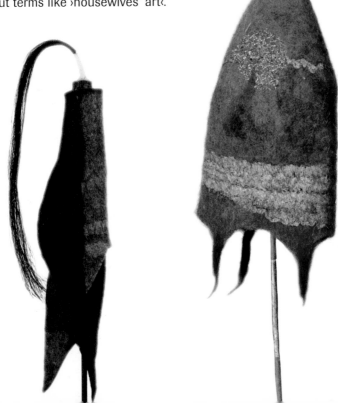

Annette Stoll Weinböhla (D)

**Spiel- und Hutobjekte I – III
mit Zipfeln zum Herausziehen und
Eindrücken, Langziehen und
Zusammenstauchen
Wollfilz
Ø ca. 18 cm, H. (zusammenge-
staucht) ca. 14 cm
1999**
Play and hat objects I – III
with corners for pulling and pressing
in, pulling out and squeezing together
Wool felt
Ø ca. 18 cm, H. (squeezed together)
ca. 14 cm
1999

Kennzeichnend für diese Spiel- und Hutobjekte sind spiralförmig eingefaltete Zipfel, die sich beliebig herausziehen und wieder hineindrücken lassen und dabei formstabil bleiben.
Die Objekte bieten vielfältige Möglichkeiten zum Spiel, wobei kein konkretes Spiel vorgegeben ist, da sie von selbst zum Anfassen und Probieren anregen. Als Kopfbedeckung können sie so auch die jeweilige Stimmung des Trägers unterstreichen.

The salient characteristic these play and hat objects share is corners folded in spiral shapes which can be pulled out at will and pushed in again, remaining stable in form all the while. The objects afford diverse possibilities for play although no particular game is prescribed since they inspire you to touch them and try them out anyway. Worn as headgear, they may also enhance the wearer's moods.

Silja Puranen Järvenpää (FIN)

Onni (Glück)
Wolle, handgefilzt, Baumwoll-
windel, Spitze, Umdruckfotografie
42 x 24 cm
1995
Onni (Happiness)
Handfelted wool, cotton baby diaper,
lace, transfer photograph
42 x 24 cm
1995

»Onni« ist in gewisser Hinsicht ein sehr persönliches Werk, sein Thema sind die ersten Lebenstage meines Sohnes Onni, der mit Down-Syndrom und einem angeborenen Herzfehler zur Welt kam. Meine Arbeit kreist bereits seit Jahren um Fragen der Weiblichkeit: der weibliche Körper – Sexualität, Mutterschaft usw. Figurative und narrative Darstellungsformen waren für diese Thematik charakteristisch. In meinen Filzarbeiten verbinde ich Wolle mit »weiblichen« Materialien wie alten Spitzen, auf den Körper bezogenen Textilien, Baumwollwindeln, Slips usw. Textilien fasse ich nicht als Rohmaterial auf, sondern als einen für das Konzept wesentlichen und von ihm nicht zu trennenden Teil der Arbeit. Ein weiteres wichtiges Element meiner neuen Arbeiten sind Fotografien (menschlicher Körper), die ich meistens digital bearbeite und in Form von Umdruckkopien auf textilem Material verwende. Die Fotografie ist für mich ein natürliches und vielseitiges Mittel, um erzählende Elemente in meine Arbeiten einzubringen.

In a way, »Onni« is a very personal work. Its subject are the early days in the life of my first son Onni, who has Down's syndrome and congenital myocardial deformation. In the past few years my work has been dealing with feminity: woman's body – sexuality, motherhood etc. Figurative and narrative expressions have been characteristic for this theme. In my felt works I combine wool with »feminine« materials like old laces, body-related household textiles, cotton baby diapers, pan clothes etc. I understand textiles not only as raw materials, but as essential and inseparable parts of a work's concept. Other important elements of my recent work have been photographs (of human bodies), which I had processed mostly digitally and used as transfer photocopies on textile material. I find photography a natural and versatile way of bringing narratives into my pieces.

István Vidák, Mari Nagy Kecskemét (H)

Filzt die Welt zusammen
Handgefilzter Teppich, Wolle
Ø 200 cm
2000
Felt the World Together
Handfelted rug, wool
Ø 200 cm
2000

István Vidák und Mari Nagy beschäftigen sich seit den frühen 80er Jahren mit Filz. Sie haben alle wichtigen Zentren der Filzherstellung bereist und verfügen selbst über eine umfangreiche Sammlung traditioneller Filzteppiche, wie sie hauptsächlich von Nomadenvölkern hergestellt werden. 1984 veranstalteten sie die Erste Internationale Filzkonferenz in Kecskemét und haben seitdem zahlreiche Ausstellungen und Workshops in Ungarn und im Ausland geleitet.

Ihre eigene künstlerische Arbeit verstehen beide auch als einen spirituellen Prozeß geistiger Reinigung.

István Vidák and Mari Nagy have been working with felt since the early 1980s. They have travelled to all the important felting centres and own a large collection of traditional felt carpets of the type made chiefly by nomadic peoples. In 1984 they staged the First International Felt Conference in Kecskemét and since then have given numerous exhibitions and held workshops both in Hungary and abroad.

They regard their own creative work as a spiritual process of intellectual cleansing and purification.

Lies Bielowski Innsbruck (A)

Filzteppich
Schafwolle, gefilzt
200 x 200 cm
1999
Felt carpet
Wool felted
200 x 200 cm
1999

Sitz- und Kopfkissen
Schafwolle und Seide, gefilzt
57 x 57 cm; 57 x 42 cm
1999
Couch and bed pillows
Wool and silk felt
57 x 57 cm; 57 x 42 cm
1999

Der Filz als Begriff im Sinne von wirklich be-greifen oder auch tun an der Sache und dadurch die Sache in der Tat finden; das Bewegen des Begriffs, sowohl auf der Ebene der Materie wie auch auf der des Denkens sind von Interesse. Bewegung in dem Sinne, daß aus dem »Chaos« über die Bewegung die Form entsteht.

Felt as a concept in the sense of really grasping or also doing something to something and thus actually finding the thing by doing; moving the concept on to both the material and the contemplative plane are of interest here. Movement in the sense that form is generated from ›chaos‹ via movement.

Laterne
Wolle
170 x 160 cm
1999
Lantern
Wool
170 x 160 cm
1999

Bernhard Früh Erfurt (D)

Ringskulpturen
Filz, Gold 750/000
Höhe 5,8–6,2 cm
1997
Auflage je 3
Ring sculptures
Felt, gold 750/000
Height 5.8–6.2 cm
1997
Editions of 3

Kay Eppi Nölke Pforzheim (D)

Vater – Mutter – Haus
9 Filsringe [sic!] mit Pflegehinweisen
Wollfilz
42 x 30 mm; 44 x 28 mm; 42 x 29 mm;
Stärke jeweils 10 mm
1999
Father – Mother – House
9 Felt rings with instructions for caring
Wool felt
42 x 30 mm; 44 x 28 mm; 42 x 29 mm;
Thickness 10 mm each
1999

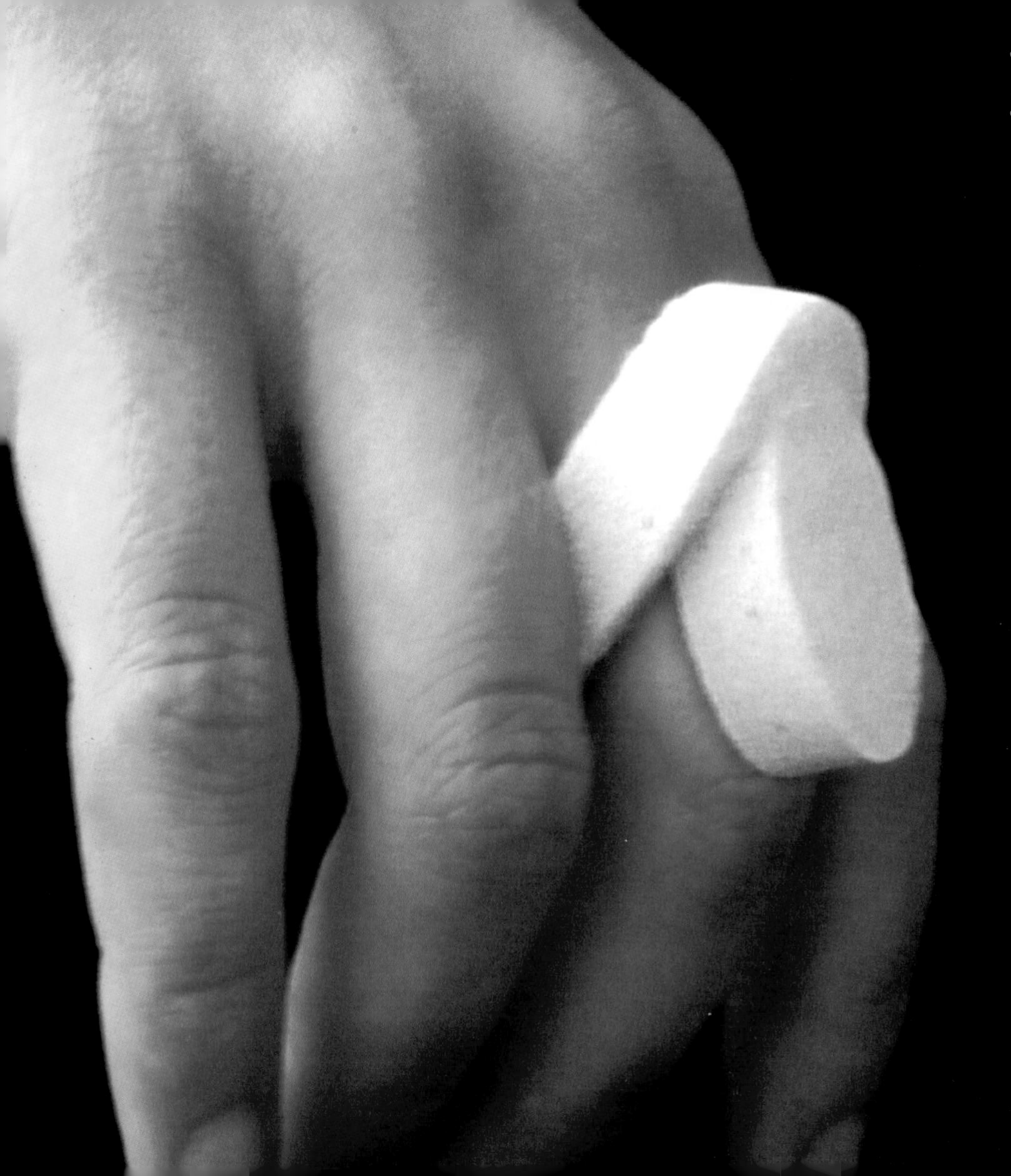

Ausgangspunkt dieser Arbeit ist die Fläche. Die separaten Pflegehinweise verweisen auf die Waschbarkeit der Ringe. Erst nach dem Lesen dieser Hinweise und dem Herausziehen der Nadel, mit der sie befestigt sind, werden die Ringe tragbar. Das zugrundeliegende gestalterische Prinzip lautet: eine Kraft wirkt, und eine Form entsteht – denn erst durch das Tragen entsteht die Form.

The starting point for this work is the plane surface. Separate instructions for caring for it point out that these rings are washable. The rings cannot be worn until these pointers have been read and the pin which fastens them has been pulled out.

The underlying design principle runs as follows: a force operates and a form is created – for form comes into being through the wearing of the rings.

Karin Wagner Basel (CH)

Blumenhalskette
Schafwolle, handgefertigt
L. ca. 80 cm
2000
Flower necklace
Sheep wool, handmade
L. approx. 80 cm
2000

Ulrich Beckert, Georg Soanca-Pollak
»the walking house« München (D)

OH! Sitzen auf Filz
Merinofilz, naturmeliert, Kanten-
streifen in 9 Farben
40 x 50 x 50 cm
Filz abnehmbar. Wahlweise mit
Tisch-Tablett, 48 x 48 cm
1999
OH! Sitting on felt
Merino felt, naturally streaked, bor-
dered edges in 9 colours
40 x 50 x 50 cm
Felt can be removed. If desired, with
a table-tray, 48 x 48 cm
1999

Milla
Sessel, Wollfilz, verstärkt, Gestell
Stahl, matt verchromt
70 x 78 x 67 cm
2000 (Prototyp)
Milla
Chair, wool felt, reinforced, steel
frame, matt chromed
70 x 78 x 67 cm
2000 (Prototype)

**Meine Arbeit beruht darauf, bestehende Typologien des Wohnens ständig weiter-
zuentwickeln, wobei ich sowohl neue als auch traditionelle Materialien in unterschiedlicher Weise verwende. Die Grundlage des Ganzen bilden Kreativität und Freiheit.
Dabei lege ich wert auf Funktionalität und Spontaneität und halte mich von wechselnden
Moden und rhetorischen Gesten möglichst fern.**
My work consists in continuing to develop further the existing typologies of living. In doing so
I use both new and traditional materials in various ways. Creativity and freedom form the
basis of the whole thing. Functionality and spontaneity are what I value most and I stay as far
away as possible from changes in fashion and rhetorical gestures.

Polku
Filzteppich
90 x 200 cm
Entwurf: Elina Huotari, Tuttu
Sillanpää
1999
Polku
Felt carpet
90 x 200 cm
Design: Elina Hutari, Tuttu Sillanpää
1999

206
207

Tessera
Teppich, Wollfilz auf synthetischem Trägermaterial
Einzelelemente 45 x 45 cm
2000
Tessera
Carpet, wool felt on a synthetic
support material
Individual elements 45 x 45 cm
2000

Dondolo
Sessel, Wollfilzbezug, abnehmbar
65 x 80 x 75 cm
1999
Entwurf Francesco Rota
Dondolo
Chair with removable cover
65 x 80 x 75 cm
1999
Design Francesco Rota

Durch ein spezielles Verfahren wird der Wollfilz mit einem synthetischen Trägermaterial fest verbunden und erhält dadurch die Form seiner Oberfläche. Der Teppich selbst besteht aus Modularelementen, die sich beliebig zusammensetzen und auseinandernehmen lassen.

The panels of the 'Tessera' carpet, which can be assembled and reassembled in wide variety of combinations, are made from a patented thermoforming procedure which fuses the virgin wool felt with innovative closed cell materials.

Anja Oehler Washington DC (USA)

Filzstuhl
Industriewollfilz, Untergestell
aus Stahl
80 x 65 x 65 cm
1995 (Prototyp)
Felt chair
Industrial wool felt, supporting frame
made of steel
80 x 65 x 65 cm
1995 (Prototype)

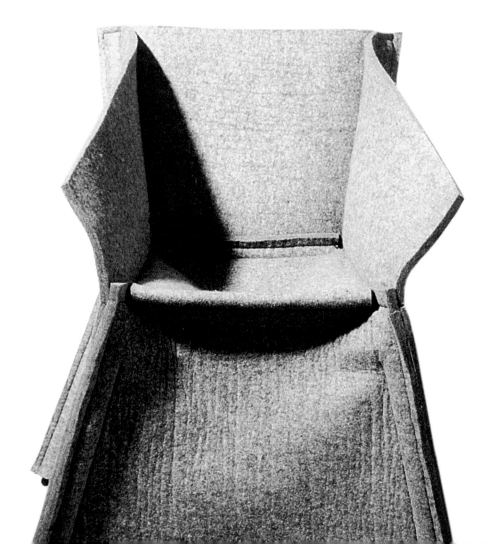

Claudia Clavuot-Merz, Ralph Feiner Chur (CH)

Liege
Walliser Schafwolle, gefilzt,
Chromstahl
L. 150 cm, B. 47 cm, H. 55 cm
1997
Lounger
Valaisan wool, felted, chrome steel
L. 150 cm, W. 47 cm, H. 55 cm
1997

Die Filzrollen nehmen – bedingt durch ihre Materialität und Ausführung – den Körper-
druck auf. Der Körper schmiegt sich in den Filz ein. Er wird in die richtige Lage, die
bequemste Paßform gebracht. Umgekehrt – durch die rollende Beweglichkeit der Filzelemente – paßt sich die ganze Liege der angewandten Kraft des Benutzers an. Die
Liege verändert ihren Standort soweit, bis sie mit der Kraft des Körpers im Gleichge-
wicht steht. Dann erst entsteht Ruhe und Bequemlichkeit. Die Filzrollen sind Stütze des
Körpers und Aufleger am Boden.
Die grundsätzliche Form des Möbels wird durch den gebogenen Stahl bewirkt. Er hält
die Rollen zusammen und zeigt – durch seine Gegensätzlichkeit – den Filz optimal in
seiner Form, Funktionalität und seinen Materialeigenschaften.

The rolls of felt take on – depending on and due to their material properties and the way the felt has been made – the pressure of your body. Your body snuggles into the felt. It is brought into the right position for the most comfortable fit. Conversely – due to the rolling movement of the felt elements – the entire lounger adapts to the force applied by the user. The lounger changes its position until it is in equilibrium with the force of your body. That is when repose and comfort set in. The rolls of felt are a support for the body and lie on the floor.

The basic form of this piece of furniture is achieved by the bent steel. It holds the rolls together and reveals – by being its exact opposite – the felt at its best in form, functionality and material properties.

Reto Kaufmann Zürich (CH)

Filzrolle
Hocker, ausgerollt als Liege
verwendbar
Wollfilz
2 Streifen 1,5 x 200 x 45 cm
1 Streifen 0,5 x 200 x 45 cm
1993 (Prototyp)
Felt roll
Stool, rolled out can be used as a
lounger
Wool felt
2 strips 1.5 x 200 x 45 cm
1 strip 0.5 x 200 x 45 cm
1993 (prototype)

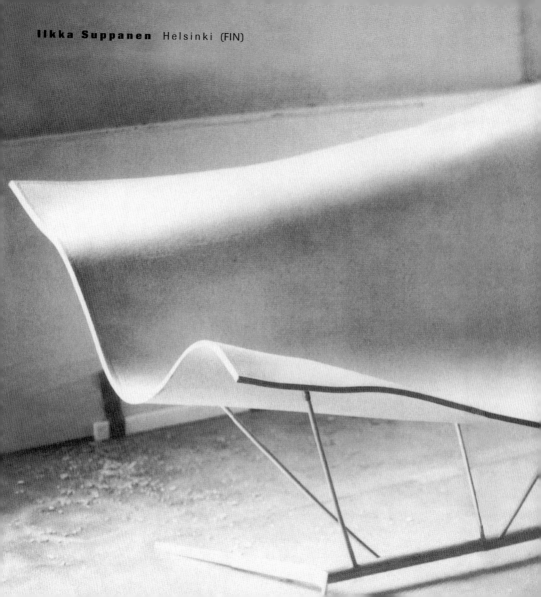

Flying Carpet
Sofa
Stahl, Filz
Hersteller Cappellini S.p.a.,
Arosio (I)
190 cm x 105 cm x 105 cm
Flying Carpet
Sofa
Steel, felt
190 cm x 105 cm x 105 cm

Timo Salli, Ilkka Suppanen, Ilkka Terho und Teppo Asikainen wurden 1997 über die Grenzen ihres Landes hinaus bekannt, als sie auf dem Salone del Mobile in Mailand mit viel Witz und Mut ihre Ausstellung »Snowcrash« zeigten. Mit erstaunlicher Leichtigkeit gelingt es ihnen, Industriematerialien wie Stahl, Glas und Kunststoff mit Federn oder Filz zu kombinieren und zugleich humorvolle Gebilde zu schaffen. [...]
Das Sofa »Flying Carpet«: Auf einem Stahlgestänge ist eine große Filzmatte befestigt, die sich sanft den Bewegungen des Sitzenden anpaßt.
(Uta Abendroth, Saarbrücker Zeitung, 28. 7. 2000)

In 1997, Timo Salli, Ikka Suppaen, Ikka Terho and Teppo Asikainen became famous far beyond the borders of their own country when, with wit and verve, they exhibited their show 'Snowcrash' at the Salone del Mobile in Milan. With astonishing lightness of touch they have succeeded in combining industrial materials like steel, glass and plastic with feathers or felt, thus creating witty objects. [...]
Take e.g. the 'Flying Carpet' sofa: a large mat of felt is fastened to a steel rod, which gently adjusts to the movements of the person sitting on it.
(Uta Abendroth, Saarbrücker Zeitung, 28 Aug. 2000)

Rolling Bag
Damentasche mit Innentasche
und abnehmbarem Tragriemen
aus Leder, Griff mit Leder unter-
legt
Wollfilz, grau-meliert
Ø 40 cm, Tiefe 11 cm
Entwurf Bernadette Ehmanns
1999
Rolling Bag
Lady's bag with pocket inside and
removable shoulder strap made out
of leather, handle lined with leather
Wool felt, mixed grey
Ø 40 cm, depth 11 cm
Design by Bernadette Ehmanns
1999

Charlotte Buch Hvalsø (DK)

Damenmantel
Merinowolle, gefilzt, Seiden-
gewebe
2000
Frock Coat
Merino wool and silk fabric
2000

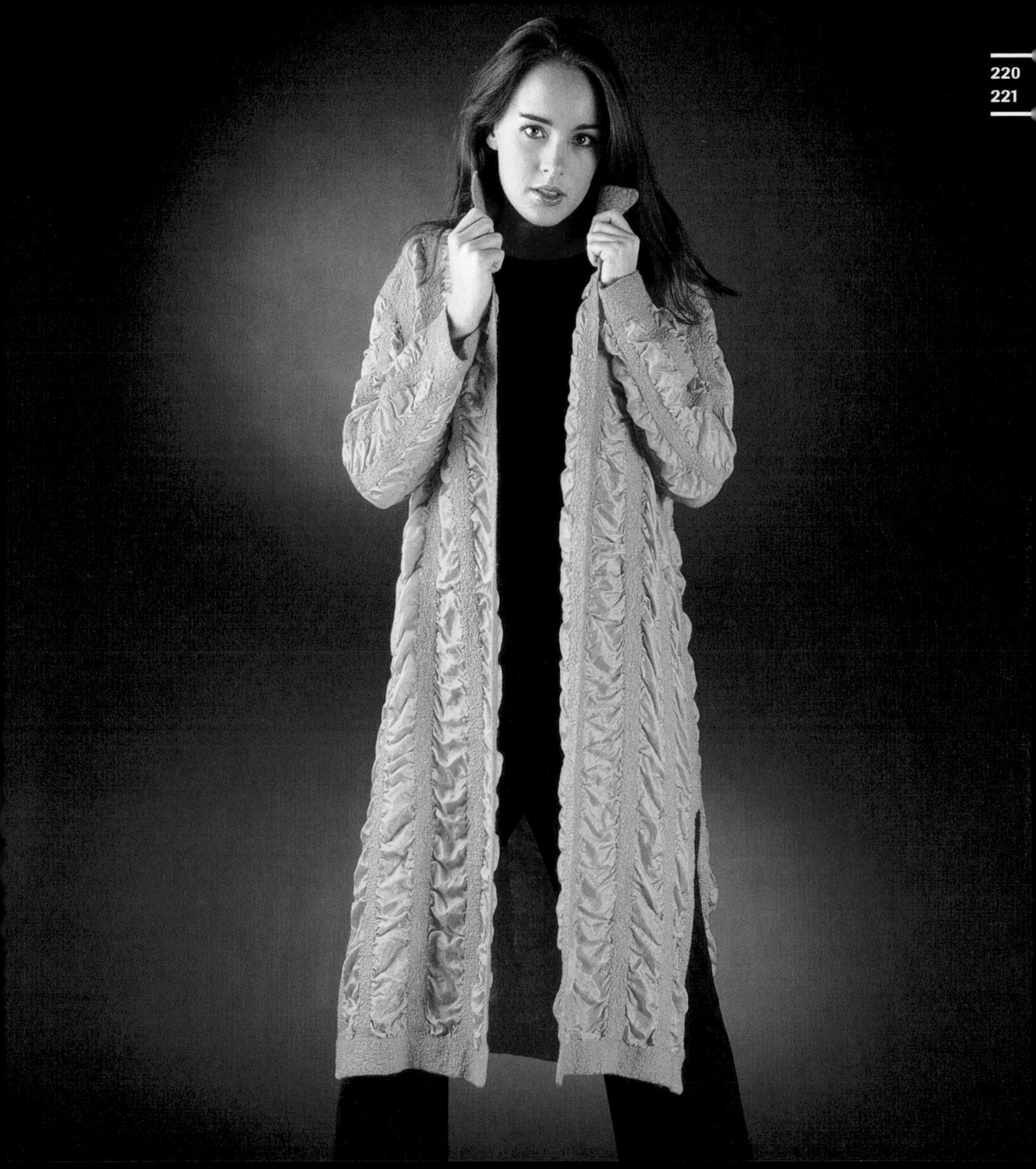

Filzcape DAVAA
aus Filzstoff Sarni Tuja FST 15
in Rot, Orange, Pink
1999
Feltscape
of Sarni Tuja FST 15 felt
in red, orange, pink
1999

Plissierter Filzstoff
FP13 AJAN
Weiß, Rot, Schwarz
140 x 200 cm
1999
Pleated felt material
FP13 AJAN
White, red, black
140 x 200 cm
2000

Ich arbeite mit Fasern, es entsteht aus
Wollfaser Filz und aus Leinenfaser Papier.
Ich arbeite mit Filz und Papier.
Daraus fertige ich Skulpturen und Reliefs,
in denen der Mensch Ausgangspunkt
und Mittelpunkt des Geschehens ist.
Für mich sind
Filz und Papier
wie
Himmel und Erde,
warm und kühl,
schwer und leicht.
Zwei Gegenpole.
Entsprechend meiner Intention setze ich beides ein.
Immer wieder habe ich mich mit der Hülle
des Menschen beschäftigt,
mit der sichtbaren wie der unsichtbaren.
Es sind Gewänder, tragbar.
Es sind Hüllen, die die Grenze zwischen
Tragbarem hin zum Objekt überschreiten.
Es sind tragbare Hüllen,
Zelt, Gewand, Teppich, Höhle in einem.
Es sind Hüllen, ausgebreitet,
sich öffnend, geschlossen, die Abfolgen
verschiedener Stadien menschlichen Seins
darstellen, Metamorphosen aufzeigen.
Es sind Elemente, die Raum erfassen und
erfaßbar machen.
Filz ebenso wie mein Papier entsteht aus
unverarbeiteten Fasern.

»Ich sehe Fasern als ein organisches Element,
welches die organische Welt
auf unserem Planeten bildet.
Sind nicht alle lebendigen Organismen aus
Faserstrukturen gebildet?
So auch der Mensch.«
(Magdalena Abakanowicz)

I work with fibres, felt is created from
wool fibres and paper from linen fibre.
I work with felt and paper.
Projects, sculptures, reliefs, materials are created
in which the human being is the be-all
and end-all of what is happening.
To me
felt and paper
are like
Heaven and Earth
warm and cool
light and heavy
Two opposite poles.
Following my intention I use both.
I have been preoccupied again and again
with the envelope of the human being
both the visible and the invisible.
These are robes, wearable ones.
These are envelopes which cross the borders between
the wearable and the object.
These are wearable envelopes,
Tent, robe, carpet, cave in one.
These are envelopes, spread out,
opening, closed, representing the sequence
of various stages of human existence
revealing metamorphoses.
These are elements which grasp space
and make space graspable.
Felt as well as my paper is created from
unprocessed fibres.

'I see fiber as the basic element
constructing the organic world
on our planet.
It is from fiber that all living
organisms are built.
We are fibrous structures.'

(Magdalena Abakanowicz)

Anna-Kaarina Kalliokoski   Isokyrö (FIN)

W.O.L. ist eine neue Modelinie, die Anna-Kaarina Kalliokoski 1999 auf der internationalen Modemesse in Helsinki vorgestellt hat.
Die luxuriösen, federleichten Filzmäntel und zarten, transparenten Schals sind aus finnischer Lambswool, Seide und Rohseide gefertigt und verbinden die jahrhundertealte Filztechnik mit modernem Design. Die Muster sind von der finnischen Natur, von Eis und verwittertem Gestein inspiriert, nehmen aber auch internationale Trends auf.

W.O.L. is a new line of clothing that Kalliokoski Design introduced to the markets at the Helsinki International Fashion Fair in 1999. Luxurious, feather-light felted coats, and fragile, transparent scarves are made of Finnish lamb wool, silk and silk fibre. The products combine a centuries-old felting technology with modern design. The ideas for the patterns come from the bare nature of Finland and from different stages of ice and rocks, but the design also includes international trends.

W.O.L. Mantel-Collection 1999
Von links nach rechts:

Pyry
Finnische Lambswool, Rohseide
Lumi
Finnische Lambswool, Rohseide
Handgefertigte Stein-Knöpfe von Riitta Piipponen.
Kuura
Finnische Lambswool, Rohseide, Seidengarn
Silber-Schließe von Kirsti Doukas.

W.O.L. coat collection 1999
From left to right:

Pyry
Finnish lamb wool, silk fibre
Lumi
Finnish lamb wool, silk fibre
Handmade rock buttons by Riitta Piipponen.
Kuura
Wool, silk fibre and silk yarn.
Silver clasp made by Kirsti Doukas.

Abendkleid
Seide, Merinowolle, Alpaca
Winterkollektion 1999/2000
Styling Christina Roth
Robe
Silk, merino wool, alpaca
1999/2000
Styling Christina Roth

Winterkollektion 1999/2000
Modenschau in Mailand
Moda Milano
Collection winter 1999/2000
Fashion show in Milan
Moda Milano

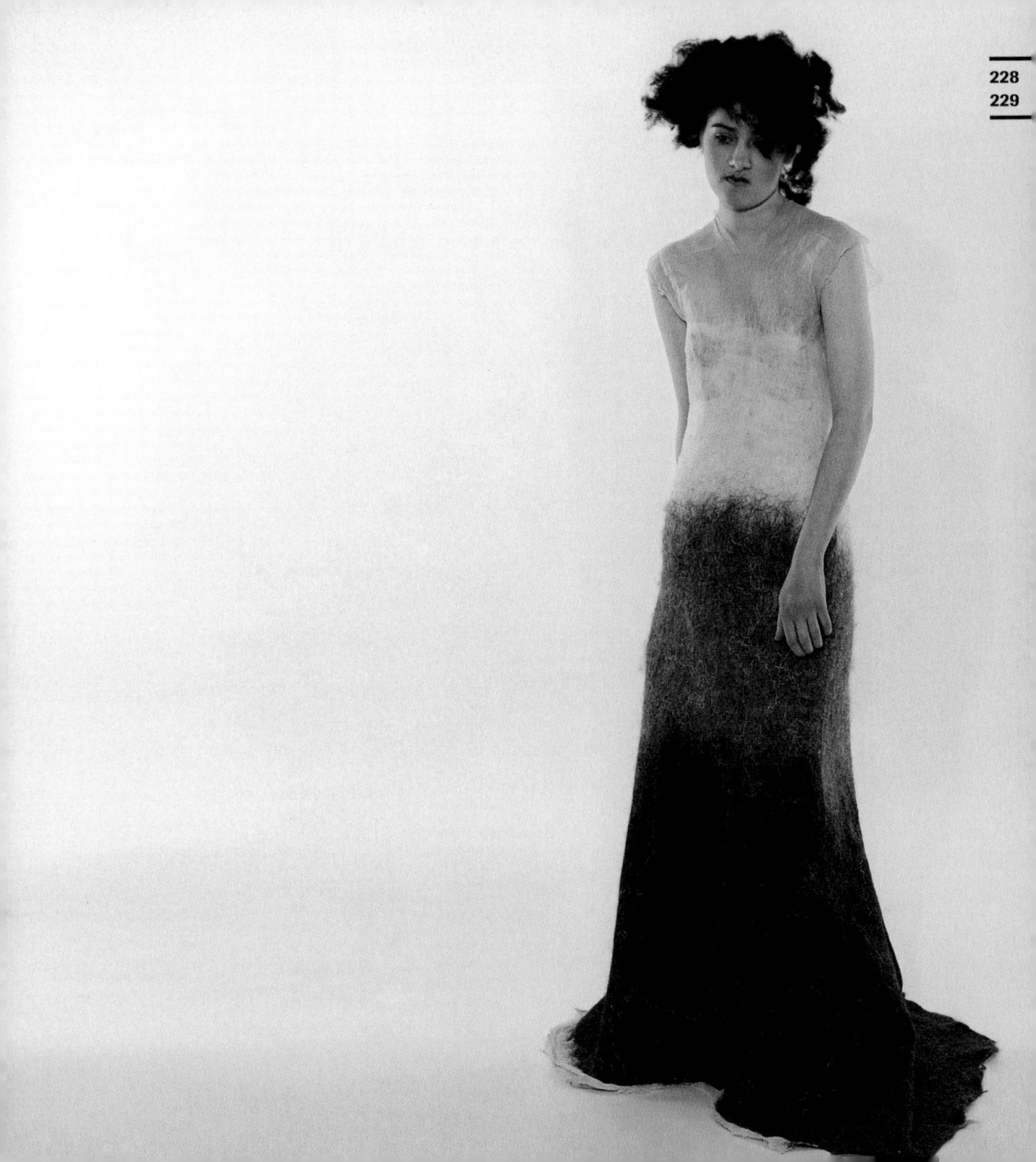

Die Renais-
sance der
Filzmacher-
kunst

Inge Evers

Einführung

Im Jahr 2000 erinnert die International Felt Association (I. F. A.) daran, daß nahezu dreißig Jahre verstrichen sind, seit Mary Burkett die Welt des Filzes wiederentdeckte und diesem wunderbaren Material zu neuer Bekanntheit und Beliebtheit verhalf. Anfangs war ihr nicht einmal klar, was es war, das sie faszinierte, als sie 1962 während einer Iranreise ihr Auto anhielt, um einen näheren Blick auf ein paar Leute zu werfen, die eine nicht definierbare Masse auf einem Laubbett hin- und herrollten. Es dauerte jedoch nicht lange, bis sie erkannte, daß sie Filz walkten. Angesichts der Ursprünglichkeit der Muster und des ebenso ungewöhnlichen wie simplen Herstellungsverfahrens fragte sie sich, wie eine so alte, traditionsreiche Kunst derart in Vergessenheit hatte geraten können.

Diese Zufallsbegegnung markierte für Burkett den Beginn eines zeit- und beschäftigungsintensiven Hobbys: Sechzehn Jahre lang widmete sie ihre Freizeit dem Thema, recherchierte und schrieb eine Geschichte des Filzes — ohne zu ahnen, daß ihre Arbeit eine ganze Generation von Filzmachern beeinflussen würde. 1971, sie leitete damals die Abbot Hall Art Gallery in Kendal (England), initiierte und organisierte Mary Burkett eine Ausstellung turkmenischer Filzarbeiten in Abbot Hall und der Whitworth Art Gallery in Manchester. 1979 dann erschien — zeitgleich mit einer gleichnamigen Wanderausstellung, die zahlreiche englische Städte besuchte — ihr Buch *The Art of the Feltmaker*. Präsentiert wurde Filz aus verschiedenen Ländern des Nahen und Mittleren Ostens sowie aus Asien und Skandinavien. Die mannigfaltigen, durchweg kraftvollen Motive der alten Textilien und Filzarbeiten waren für die Öffentlichkeit die reinste Offenbarung. Filz spricht die Sinne an. Er lädt zum Anfassen ein, zum mit ihm und auf ihm Leben. Manche Besucher kamen nur, um den Geruch zu schnuppern. Künstler, Kunsthandwerker, Studenten und Dozenten waren gleichermaßen überrascht und gebannt. Und Ausstellung und Buch lösten eine regelrechte Welle der Begeisterung aus. Mary Burkett förderte unser Wissen hinsichtlich der Geschichte des Filzens weiter – durch weite (Vortrags-)Reisen und Veröffentlichung neugewonnener Erkenntnisse in der 1986 gegründeten Zeitschrift Echoes, dem Organ der »Filzgemeinde«. Für ihre Museumsarbeit wurde sie mit dem Order of the British Empire und einem Fellowship der Museum Association ausgezeichnet, und Filzkünstler verliehen ihr den liebevollen Ehrentitel *MOF*, »Mother of Felt«.

Die Renaissance der Filzmacherkunst

Seit Ende der 60er Jahre beschäftigten sich europäische und amerikanische Textilkünstler mit den unterschiedlichsten Aspekten von Faserkunst und erkundeten in diesem Zusammenhang auch die Ursprünge verschiedener Werkstoffe. Das Filzen entwickelte sich beim Experimentieren mit Fasermaterialien. Die nachfolgend aufgeführten Pioniere und Pädagogen, Künstler und Autoren stehen als Repräsentanten für das bislang Erreichte.
1976 veröffentlichte Katharina Ågren im schwedischen Västerås ihr Buch *Tovning* (»Filzen«), das ins Englische und sogar ins Niederländi-

sche übersetzt wurde. In den Vereinigten Staaten kam 1980 Beverly Gordons *Feltmaking* heraus, das Geschichte und damaligen Stand des Filzmacherhandwerks beleuchtete und Schritt-für-Schritt-Anleitungen zum Handfilzen enthielt. Abbildungen zeigten Werke von Künstlern wie Beth Beede sowie Joan Livingstone, die bereits seit den späten sechziger Jahren Filz herstellten. Beide hielten Vorträge und hatten zahlreiche Ausstellungen. So waren natürlich auch beide vertreten, als das New Yorker American Craft Museum 1980 unter dem Titel *Felting* traditionelle und zeitgenössische Filzarbeiten präsentierte.

Anfänge eines internationalen Netzwerks

Zum Gedenken an Veronica Gervers, einer Pionierin auf dem Gebiet der Geschichte der Textil- und insbesondere der Filzkunst, organisierte István Vidák 1984 im ungarischen Kecskemét die First International Felt Conference (Erste Internationale Filzkonferenz) für Ethnographen, Architekten, Textilhistoriker und Filzmacher. Im Garten des Szórakaténusz-Museums war eine Reihe von Jurten aufgestellt, und alte Filzarbeiten aus der Sammlung von István Vidák und seiner Frau Mari Nagy schmückten die Museumswände. Mary Burkett hielt einen Vortrag über die Geschichte des Filzes, David Tschitschiwili aus Georgien referierte über traditionelle Filzherstellung der Tusheti, Beth Beede aus den USA über Filzfabrikation in Amerika, Bill Copperthwaite (USA) und Peter Andrews, der aus England stammt, aber in Deutschland lebt und arbeitet, über Filzarchitektur. Edith Hollos aus Ungarn sprach über Schamanenforschung und Filzarbeiten, und die Holländerin Inge

Evers über Filz als Gemeinschaftskunst im »Haarlem-Felt-and-Paper-Workshop«. Deutschland wurde von Katharina Thomas vertreten, einer besonders aktiven Pionierin des Fachbereichs.

Die Konferenz von Kecskemét erweiterte den Horizont der Filzforschung, schlug Brücken zwischen den einzelnen Disziplinen und legte den Grundstein für die künftige internationale Zusammenarbeit. In den folgenden vier Jahren organisierte Vidák internationale Meetings, zu denen auch Meister aus Zentralasien eingeladen waren, um ihr traditionelles Wissen an Filzkünstler aus dem Westen weiterzugeben. Die Teilnehmer kamen aus England, Schweden, Österreich, Dänemark, Ost- und Westdeutschland, Frankreich, Australien, Neuseeland, Japan und den Vereinigten Staaten. Wer sich regelmäßig an den Treffen dieser Jahre beteiligte, kam in den Genuß einer exklusiven Schulung und enormen Hintergrundwissens.

Ein Verband, ein Newsletter, ein Konvent und ein Filzmuseum

In diesen Pioniertagen hatten die Filzmacher noch keine eigene Organisation. Engagierte Männer und Frauen wie Beatrijs Sterk in Hannover, die 1982 die erste Nummer ihrer Zeitschrift *Textilforum* dem Thema Filz widmete, verhalfen der Renaissance des Filzens jedoch zu weiterer Bekanntheit. Englische Filzkünstler gründeten 1984, dem Jahr der Ersten Internationale Filzkonferenz, mit der »Feltmakers' Association« den ersten Filzmacherverband. Die Ausstellung *Art of the Feltmaker* hatte eine Reihe von Künstlern und Dozenten — darunter Eva Kuniczak, Jenny Cowern, Bailey Curtis, Freda Walker und Peter Walter von Hat Industry Bury Cooper Whitehead Ltd. – angeregt, sich auf das Abenteuer Filz einzulassen, was letztlich zur Gründung dieses Verbandes führte. Zu den Gründungsmitgliedern zählten Annie Sherburne, die erste Filzkünstlerin, die mit Industrie und Modemachern zusammenarbeitete, Pippa Lewis und Victoria Brown. Einige Jahre gab Lene Nielsen die Zeitschrift *Felt Filt* heraus, die neben vielem anderen auch Arbeitsproben enthielt.

1992 gründete Pat Spark den *North American Felters' Newsletter*, seither Sprachrohr der nordamerikanischen Filzgemeinde. Spark betreut das Blatt bis heute als Herausgeberin. Das erste Treffen der Filzmacher hatte 1978 im Rahmen eines Konvents in Fort Collins, Colorado, stattgefunden. Chad Alice Hagen organisierte mit dem '94er Konvent in Minneapolis, Minnesota, eine zweite Zusammenkunft. Pat Spark setzte die Tradition 1996 in Portland fort, und der Konvent von 1998 in Atlanta, Georgia, umfaßte erstmals auch informelle Meetings. Die erste internationale Filzkonferenz in Nordamerika wurde 1994 im National Exhibit Center in Castlegar, British Columbia, Kanada, abgehalten, die nächste fand 1997 im Lost Valley Education/Retreat Center unweit Eugene, Oregon, statt. Die dritte Konferenz dieser Art wurde schließlich im Rahmen des »Michigan Fiber Fest« im Jahre 2000 auf dem Messegelände von Allegan County in Michigan veranstaltet.

Weit entfernt von diesem Geschehen, jenseits des großen Teiches, öffnete 1987 im nordostfranzösischen Mouzon ein Filzmuseum seine Pforten. Mit diesem »Musée du Feutre« schufen Paul Motte, Richard

Keller, Agnes Paris und Serge Chaumier eine ganz den kulturellen, wissenschaftlichen, technischen und industriellen Aspekten von Filz gewidmete Institution. Im selben Jahr wurde Mary Burkett Präsidentin der »English Felt Association«, die 1988 in »International Feltmakers' Association« umbenannt wurde.

Filzfestivals und vieles mehr

Diese und andere Treffen der Filzgemeinde setzten einen kontinuierlichen Strom des Lernens und Austausches in Gang. Immer mehr Filzmacher organisierten sich, und in zahlreichen europäischen Städten fanden Symposien und Ausstellungen statt. Hier eine Auswahl: In Korros (Schweden) richtete Marianne Ekert 1990 eine Filzkonferenz mit angeschlossenem Festival aus. Im gleichen Jahr veranstalteten Annette Damgaard und Lene Nielsen im dänischen Aarhus eine internationale Konferenz und Ausstellung. Ein Jahr später riefen die holländischen Filzmacher Corrie Balk und Marion Jacobs den »Viltkontakt« ins Leben. Und 1994 feierte die International Felt Association ihr zehnjähriges Bestehen mit einer Tagung in Hartpury House, Gloucester (England). Elisabeth Weissensteiner aus Österreich sprach über die mythologische Bedeutung von Textilien, Jorie Johnson kam mit einer Gruppe aus Japan, Layne Goldsmith (USA) hielt einen Vortrag, desgleichen Nino Kipschidse aus Georgien, die über Tusheti-Filze sowie traditionelle Handwerkstechniken und Muster aus Tiflis referierte. Inzwischen war die Organisation international geworden. Lene Nielsen (Dänemark), Katharina Thomas (Deutschland), Kathie

Hoppler-Dinkel (Schweiz) und Jenny Mackay (Schottland) lieferten Beiträge, ebenso Jenny Cowern, Jeanette Appleton, Stephanie Bunn und Janet Ledsham (alle England). Im Jahr darauf organisierte Jette Clover mit Inge Evers als Beraterin im Textilmuseum des holländischen Tilburg eine Ausstellung und Konferenz – komplett mit Workshops und Schafschurvorführung. Mary Burkett referierte über Joseph Beuys, und Barbara Tietze (Deutschland) äußerte sich zum Thema Ergonomie und Filz.

Im Sommer des Jahres 1994 fand in der Collins Gallery im schottischen Glasgow die internationale Ausstellung, Workshop und Konferenz »Felt Directions« statt. Zwei Jahre später, 1996, luden drei schweizer Filzmacherinnen, Verena Gloor und die inzwischen leider beide verstorbenen Leni Hunger und Helen Widder-Maeschi, zu einem interantionalen Filzmachersymposion nach Landquart in der Nähe von Chur. Das breite Spektrum der Vorträge reichte von dendrologischen bis hin zu theologischen und psychologischen Themen.

Bei der kurz zuvor abgehaltenen Jahrestagung der I.F.A. hatte Stephanie Bunn ausführlich über Filzherstellung in Kirgisien berichtet – das Thema ihrer Doktorarbeit im Fachbereich Anthropologie. Sie wurde 1996 Bildungsbeauftragte des Verbands.

Bereits 1995 hatte Lesley Blythe-Lord den Posten des Herausgebers von *Echoes* übernommen, und die für ihre Hüte bekannte Meike Dalal-Laurenson zeichnet seit demselben Jahr für den Bereich Ausstellungen verantwortlich.

Australien und Neuseeland haben in den vergangenen zwei Jahrzehnten ebenfalls beachtliche Fortschritte erzielt, einerseits mit hochinteressanten Experimenten und zum anderen durch die Zusammenführung internationaler Einflüsse. Das »Australian Forum for Textile Arts« veranstaltet nicht nur internationale Textiltagungen, sondern veröffentlicht zudem das Journal *Textile Fibre Forum*, als dessen Herausgeberin seit zehn Jahren Janet de Boer fungiert, die auch die Tagungen organisiert. In vielen australischen Bundesstaaten existieren Filzmachergruppen, unter denen als aktivste die »Victorian Feltmakers« und die »Canberra Region Feltmakers« hervorstechen, die eigenständig Ausstellungen, Workshops, Seminare und Konferenzen abhalten. Um die nationale Zusammenarbeit trotz der riesigen Entfernungen des Fünften Kontinents zu erleichtern, gründete die Südaustralierin India Flint vor kurzem den *Australian Feltmakers' Newsletter*. »Southern Hemisphere Felters' Workshops« fanden zweimal in Neuseeland und einmal, 1994, in Bunbury, Westaustralien, statt.

Filzkünstler und Nomaden

Das im Sommer 1999 in Finnland abgehaltene internationale Filzsymposion »The Wandering of the Midnight Sun« (Der Weg der Mitternachtssonne) wurde von Leena Sipila organisiert, die dem Direktorium des »Central Finland Arts and Crafts Institute« (Zentralfinnisches Institut für Kunst- und Kunsthandwerk) angehört. Wie unzählige Male zuvor hatte man Mary Burkett gebeten, die Eröffnungsrede zu übernehmen und auch einen Vortrag zu halten. Das Programm der Veranstaltung las sich wie ein Who-is-who der Filzgemeinde und spiegelte die professionalisierte Entwicklung von Filzkunst und -handwerk seit 1975. Die von den vier finnischen Tutoren zur Diskussion gestellten Themen reichten von mongolischem Pferdefilz, präsentiert von Tuula Nikulainen, bis zu »Filzwanderungen« von Marjo Tuhkanen, »Erde, Kunst und Filz« von Cari Caven und Fotografien auf Filz von Silja Puranen. Mari Nagy (Ungarn) führte für alle Interessenten offene Workshops durch, und Katharina Thomas und Agnes Keil (beide Deutschland) leiteten den Musik-Tanz-Filz-Workshop »RED«. Inge Evers (Holland) setzte in ihrem Workshop zum Thema Filz als »sinnliche« Erfahrung eine breite Palette unterschiedlicher Materialien ein, während Gunilla Paeteau (Schweden) »Kinder und Filz« präsentierte, wozu unter anderem eine von Engelsfiguren aus dem örtlichen Gotteshaus inspirierte Gruppenarbeit gehörte. Weitere Projekte waren Jorie Johnsons (Japan) »Filz-Flächen-Experimente«, »Dekorationen auf Filz« von Patricia Spark (USA), »Stoff-Filz« von Alexander Pilin (Georgien), der darüber hinaus eine Modenschau inszenierte, und »Semi-industrieller Filz« von István Vidák (Ungarn), das den Einsatz von Filz in der Industrie, aber auch auf den Gebieten Kunst, Mode und Interior Design illustrierte.

Der Kreis schließt sich

Seit 1979 hat *The Art of the Feltmaker* den Erdball umrundet. Heute dient Filz wieder einer Vielzahl von Verwendungszwecken, in der Bekleidungsindustrie und Modebranche ebenso wie in der modernen Kunst und für praktischen und »therapeutischen« Gebrauch. Im Augenblick erlebt der vielseitige Werkstoff einen regelrechten Boom; selbst Filme wie *Star Wars* setzen Filzkostüme ein. Mary Burketts Traum von einer Filz-Renaissance ist wahr geworden. Filzmacher des zwanzigsten Jahrhunderts sind Pioniere und Wiederbeleber in einem. Auch hinsichtlich der Herstellungsverfahren wurde Neuland betreten. Nomaden gleich, reisen Filzkünstler um die halbe Welt, um ihr Wissen an die nächste Generation weiterzugeben. Für einige jedoch ist es Zeit, sich niederzulassen. 1999 gründete Lene Nielsen an Skals Handarbejdsskole die erste dänische Filzmacherakademie. Und im Januar 2000 fand gleichfalls in Dänemark eine internationale Minikonferenz statt, auf der Studenten und professionelle Filzmacher neue Techniken und Materialien kennenlernen, zeitgenössische Handwerker, Designer und Künstler treffen und mehr über die Kulturgeschichte des Filzes erfahren konnten.

Präsenz zeigt Filz auch in St. Petersburg, wo 1999 beim Vierten Internationalen »White Nights«-Textil-Symposium Filzkunst und -handwerk gezeigt wurden. Im gleichen Jahr richtete Tiflis mit der »Caucasian Textile Route« das Zweite Europäische Textil-Netzwerk aus. Dazu gehörten Vorträge und Ausstellungen — und natürlich georgische Künstler wie Kipschidse, die beide Events organisiert hatte, Nino

Tschatschkjani, T. und A. Sisauri, Maia Zinamdschurischwili, Nino Kwawilaschwili, Kethij Kawtaradse, Tamara Gelaschwili, die aus Georgien gebürtige Petersburgerin Tina Kutalja und last not least der legendäre Tengis Japaridse, der bereits seit Ende der sechziger Jahre Filzbilder kreiert. Mary Burkett, Helena Selinenkowa (eine Ethnographin aus St. Petersburg), Mari Nagy und Inge Evers hielten Vorträge und/oder Workshops, außerdem referierte Jeanette Appleton über Filz und Kunst und Pat Spark über Filzmacher in den USA.

Comeback mit Visionen

Das Jahr 2000 ist ein Gedenkjahr. Beim Filzfestival in Kendal wurde Mary Burkett als Symbol der immensen Wertschätzung, die man ihr heute weltweit entgegenbringt, ein »Mille-Fleurs«-Filzteppich überreicht. Ende April entwarf und fertigte ein Team von sechsundzwanzig erfahrenen Filzmachern aus fünf Ländern innerhalb vier Tagen dieses zwei mal drei Meter große Stück, in das 250 Filzblumen integriert wurden, die Filzmacher aus aller Welt zu eben diesem Zweck geschickt hatten. Das unter Inge Evers' Regie von Sheila Smith koordinierte Millenniumsprojekt zeigte darüber hinaus aber auch neue Richtungen für »Filz als Gemeinschaftskunst« und »Filz als sinnliches Erlebnis« auf.

Heute kommunizieren Filzmacher aus aller Welt via Internet und treffen sich als Computernomaden im Cyberspace. Ist ein größerer Unterschied zu jenen Menschen, »die eine nicht definierbare Masse auf einem Laubbett hin- und herrollten«, überhaupt denkbar? Frühe Filzkünstler signierten ihre Arbeiten nicht und werden für immer unbekannt bleiben. Ihrer Gemeinschaft zu dienen, war ihnen Lebenszweck. Was aber hat ein anonymer Filzmacher im Mittleren Osten, der in seiner Jurte auf einem Filzteppich liegt und durch das runde Deckenloch starrt, mit einem jungen Filzmodedesigner gemein, der in der westlichen Welt einen großen Namen trägt? Die Antwort ist einfach: Beiden scheinen nach oben keine Grenzen gesetzt und die Sterne sind zum Greifen nah.

Teppich aus Turkmenistan
Carpet from Turkmenistan

Beth Beede (USA)
»Ohne Titel«, Filzarbeit, 1993
'Untitled', felt work, 1993

Katharina Thomas (D)
»Filzhochzeit«, Preis der Textil-
kunstbiennale, Krefeld 1987
'Felt Wedding', prize at the Textile
Art Biennial, Krefeld 1987

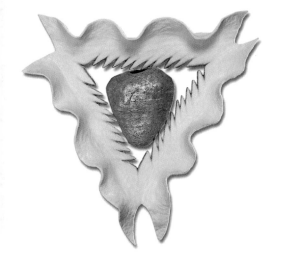

Layne Goldsmith (USA)
»Fear Not«, Mischtechnik auf Wollfilz, 1992
'Fear Not', mixed media on wool felt, 1992

Mollie Littlejohn (AUS)
'Wall Mattress'

Chad Alice Hagen (USA)
'Driveway Collection 3', Detail, 2000

Annie Sherburne (GB)
Filzarbeit
Felt work

Jeanette Appleton (GB)
'Tidal Flow II', 1999

The Revival of the Feltmaker's Art

Inge Evers

Introduction

The year 2000 has been chosen by the International Felt Association to commemorate the fact that almost thirty years have passed since Mary Burkett brought new light and life into the world of felt. At first she did not even know it was felt what fascinated her when on a trip to Iran in 1962, she stopped her car for a closer look at people who were rolling a bundle of something on a bed of leaves. It did not take her long to realize it was felt they were making. Recognizing the antiquity of the designs and the unusual and simple nature of the process, she wondered why such an ancient art was close to extinction.

This private encounter was the beginning of a 16-year period during which in her spare time she researched and wrote a history of felt, without an inkling of the impact this work would have on a whole generation of feltmakers. In 1971, when she was the director of the Abbot Hall Art Gallery in Kendal (England), Mary Burkett initiated and organized an exhibition of Turcoman felt and tribal art at Abbot Hall and the Whitworth Art Gallery in Manchester. In 1979 at Abbot Hall, her book *The Art of the Feltmaker* was published and an exhibition held under the same title. It showed felts from many Middle Eastern countries as well as from Asia and Scandinavia. The exhibition toured all over England. Experiencing the rich visual imagery of ancient textiles and felts was a revelation to the public. Felt speaks to the senses. It invites to be touched, to be lived in and on. Some visitors just came for the smell of it. Artists, craftsmen, students and members of schoolboards were amazed and delighted. The exhibition and the book worked as a catalyst. For her museum work Mary Burkett was honoured with an

OBE (Order of the British Empire); she is also a FMA (Fellow, Museum Association) and has an MA. Feltmakers celebrate her as the MOF, 'Mother of Felt'.

The revival of the feltmaker's art

Since the late 1960s, textile artists in Europe and North America have been exploring all aspects of fibre arts, including the origins of various fabrics. Feltmaking developed through experimenting with fibrous materials. The pioneers, educators, artists and authors mentioned below are representative of what has so far been achieved.

In 1976, Katharina Ågren published her book *Tovning* (Felting) in Västerås (Sweden), which was translated into English and even Dutch. In the United States, Beverly Gordon's book *Feltmaking* came out in 1980; it covered the history and contemporary situation of felt-making and gave step-by-step instructions for making felt oneself. She included works of artists like Beth Beede and Joan Livingstone, the latter a feltmaker since the late sixties. Both exhibited and lectured widely. Their work was also shown at the exhibition of traditional and contemporary work mounted in 1980 at the American Craft Museum in New York under the title *Felting*.

Start of an international network

In remembrance of Veronica Gervers, one of the pioneer historians of textile art and specially felt, István Vidák in 1984 organized the First International Felt Conference at Kecskemét in Hungary. He brought together ethnographers, architects, textile historians and feltmakers. A collection of yurts (or more correctly *ger*, because originally the building was called *ger*, and the ground on which it stood was the *yurt*) was on show in the garden of the Szórakaténusz Museum, and old felts collected by István Vidák and his wife Mari Nagy covered the museum's walls. Mary Burkett lectured on the history of felt, David Tsitsishvili from Georgia on traditional feltmaking by a group of women in Tusheti, Beth Beede from the United States on feltmaking in America, Bill Copperthwaite (USA) and Peter Andrews from England, but living and working in Germany, on felt architecture, Edith Hollos from Hungary on her shaman research and feltworks, and Inge Evers from Holland on felt as community art in the Haarlem Felt-and-Paper Workshop. Katharina Thomas, one of the most active pioneers in the field, represented Germany.

This conference widened the horizon of felt studies, acted as a kind of border crossing and laid the foundations of an international network. During the next four years, Vidák organized international meetings, inviting masters from the East to share their knowledge with Western feltmakers, with participants from Ireland, Great Britain, Sweden, Austria, Denmark, both Germanies, France, Australia, New Zealand, Japan, the USA and Canada. During these years, regular participants received an exclusive training and insight.

An association, a newsletter, a convergence and a felt museum

During the pioneering days, feltmakers did not have their own organization. Thanks to editors like Beatrijs Sterk in Hannover (Germany), who published a special issue of her journal *Textilforum* on felt in 1982, news spread about the revival of felt and its makers. English felters started organizing themselves in 1984, the year the First International Felt Conference took place, and founded the Feltmakers' Association. The *Art of the Feltmaker* exhibition had propelled a number of artists and teachers – among them Eva Kuniczak, Jenny Cowern, Bailey Curtis, Freda Walker and Peter Walter from the Hat Industry Bury Cooper Whitehead Ltd – on an adventure into felt that lead to the founding of this association. Among its founding members were Annie Sherburne, the first felt artist who worked with industry and fashion designers, and Pippa Lewis and Victoria Brown. For several years, Lene Nielsen published the journal *Felt Filt*, which included samples and much else.

In 1992, Pat Spark started the *North American Felters' Newsletter*, which functions as a network for North America, and has been editing it ever since. The first meeting of felters was in 1978, at the Convergence in Fort Collins, Colorado. Chad Alice Hagen and colleagues organized the second gathering at the '94 Convergence in Minneapolis, Minnesota. Pat Spark carried on this tradition at the '96 Convergence in Portland, Oregon. At the '98 Convergence in Atlanta, Georgia, informal meetings took place. The first international felt conference held in North America took place in 1994 at the National Exhibit Center in Castlegar, British Columbia, Canada. The second was held in 1997 at the Lost Valley Education/Retreat Center near Eugene in Oregon, and the third conference was held in 2000 during the »Michigan Fiber Fest« in the Allegan County Fairgrounds in Michigan.

Far away from all this, in 1987 a felt museum, the 'Musée du Feutre', was founded in Mouzon in north-east France. There Paul Motte, Richard Keller, Agnes Paris and Serge Chaumier created a centre devoted to the cultural, scientific, technical and industrial aspects of felt. In England in the same year, Mary Burkett became President of the English Felt Association, which was renamed International Feltmakers' Association in 1988.

An unbroken string ...

Like an unbroken string of beads, a continuous stream of learning, sharing and exchange had now started, and feltmakers were organizing themselves. Symposia and exhibitions took place in various European countries. Here is a selection.

A felt conference and festival was organized by Marianne Ekert in Korros (Sweden) in 1990. In the same year, Annette Damgaard and Lene Nielsen put together an international conference and exhibition in Århus (Denmark). One year later, the Dutch feltmakers Corrie Balk and Marion Jacobs started 'Viltkontakt' to organize two meetings every year for feltmakers to exchange experiences. In 1994, the International Felt Association held a conference at Hartpury House, Gloucester (England), to celebrate its tenth anniversary. There were talks from Elisabeth Weissensteiner from Austria on the mythological implications of textiles, Jorie Johnson brought over a group from Japan, Layne Goldsmith (USA) gave a talk, as did Nino Kipshidze from Georgia, who spoke about the Tusheti felts and traditional methods and designs from Tblisi. By then the organization had become international. Lene Nielsen (Denmark), Katharina Thomas (Germany), Kathie Hoppler-Dinkel (Switzerland), Jenny Mackay (Scotland), and Jenny Cowern, Jeanette Appleton, Stephanie Bunn and Janet Ledsham (all from England) contributed. In Holland later that year, Jette Clover with Inge Evers as consultant organized an exhibition and a conference as well as workshops and a sheep-shearing demonstration at the the Dutch Textile Museum in Tilburg. Mary Burkett lectured on Joseph Beuys, while Barbara Tietze from Germany spoke about felt and ergonomics.

In the summer of 1994, an international felt exhibition, workshop and conference entitled "Felt Directions" was held at the Collins Gallery in Glasgow. In the summer of 1996, another international feltmaking symposium was hosted by the Swiss feltmakers Verena Gloor, the late Leni Hunger, and the late Helen Widder-Maeschi at Landquart near Chur. The lectures covered a wide range of subjects, including dendrology, theology, psychology and chromatics.

At the I.F.A.'s annual general meeting earlier that year, Stephanie Bunn focused on feltmaking in Kirgizya, the subject of her Ph.D. degree in anthropology. She became education officer in 1996.

When the editor's seat of *Echoes* fell vacant in 1995, it was filled by Lesley Blythe-Lord. Meike Dalal-Laurenson, known for her hats, became exhibition officer.

Australia and New Zealand have made great advances in the past two decades, with much meaningful experimentation as well as the incorporation of international influences, particularly through tutors invited to international textile forums organized by the 'Australian Forum for Textile Arts', which also publishes the Australian journal *Textile Fibre Forum*. Janet de Boer has been organizing these forums for the past ten years; she also edits the journal. A number of felting groups exist in various Australian states, two of the most lively being the Victorian Feltmakers and the Canberra Region Feltmakers, which actively promote exhibitions, offer workshops and hold conferences and retreats. India Flint from South Australia recently began publishing a national *Australian Feltmakers' Newsletter* as a way to improve networking across Australia's huge distances. The Southern Hemisphere Felters' Workshops were twice held in New Zealand and later once in Bunbury, Western Australia, in 1994.

Felt masters, nomad travellers

The International Felt Symposium, 'The Wandering of the Midnight Sun', held in central Finland in the summer of 1999, was organized by Leena Sipila, one of the directors of the Central Finland Arts and Crafts Institute. As so often before, Mary Burkett was asked to open the proceedings and hold a lecture. The program read like the contents page of a book on felt and reflected the professional development of felt art, crafts and studies since 1975. Subjects ranged from Mongolian horse felt presented by Tuula Nikulainen, to felt wandering by Marjo Tuhkanen, earth, art and felt by Cari Caven and photographs on felts by Silja Puranen. Mari Nagy (Hungary) taught workshops open to the public, and Katharina Thomas and Agnes Keil (both from Germany) led the music-dance-felt workshop 'RED'. Inge Evers (Holland) in her workshop used a wide range of natural materials on felt as a sensitive background, and Gunilla Paeteau (Sweden) presented 'Children and Felt', which, among others, included a group work inspired by figures of angels from the local church, 'Felt Surface Experiments' by Jorie Johnson (Japan), 'Decorations on Felt' by Patricia Spark (USA), 'Cloth Felt' by Alexander Pilin (Georgia), who also presented a fashion show, and 'Semi-Industrial Felt' by István Vidák (Hungary), illustrating how felt is being adopted by the industry and also used in art, fashion and interior design completed the programme.

A full cycle

Since 1979, *The Art of the Feltmaker* has made a full cycle around the globe. Today, felt is found in a broad range of applications, in dress and fashion, in modern art as well as for practical and therapeutic use. At the moment, it is highly fashionable; films like *Star Wars* make use of felt costumes. Mary Burkett's dreams about felt have come true. Felters in the twentieth century are pioneers as well as revivers.

Much new ground has been broken in terms of technique. Felters travel the world like nomads to teach and pass on their knowledge to a younger generation. For some of them it is time to settle down. In 1999, Lene Nielsen founded the first Danish school of felting at the Skals Handarbejdsskole (school of handicrafts in Skal). To teach and learn techniques and materials and, among other things, the cultural history of felt as well as to meet contemporary craftsmen, designers and artists, an international mini-conference for students and professional feltmakers was held in Denmark in January 2000.

Felt has continued to be present in St Petersburg, where felt arts and crafts were shown during the Fourth International 'White Nights' Textile Symposium in 1999. In the same year, Tblisi staged the Second European Textile Network 'Caucasian Textile Route'. There were lectures and exhibitions and also Georgian artists like Nino Kipshidze – who had organized the event for the second time – Nino Chachkiani, T. and A. Sisauri, Maia Tsinamdzgvarishvili, Nino Kvavilashvili, Kethy Kavtaradze, Tamara Gelashvili, Tina Kutalia from Petersburg but born in Georgia and, last not least, the legendary Tengiz Japaridze who has been making pictorial felts since the late 1960s. Lectures and/or workshops were held by Mary Burkett, Helena Selinenkova, an ethnographer from Petersburg, Mari Nagy and Inge Evers, where Jeanette Appleton talked about felt and art, and Pat Spark on feltmaking in the USA.

The sky is the limit

The year 2000 is a year of commemoration. During the 'Festival of Felt' in Kendal, a 'Mille Fleurs' felt carpet was presented to Mary Burkett, as a token of the regard in which she is held by feltmakers worldwide. Over a period of four days at the end of April, a team of 26 experienced feltmakers from five different countries designed and made a felt carpet of 2 x 3 metres, using 250 refelted flowers sent by feltmakers from all over the world. This millennium project, co-ordinated by Sheila Smith and directed by Inge Evers, also emphasized new directions in 'felt as community art' and 'felt for the senses'.

Feltmakers from all over the world are networking on websites, computer nomads meeting in cyberspace. How different are they compared to the felters of old rolling their 'bundles of something'? Earlier feltmakers did not sign their work and remained shrouded in anonymity. Serving their communities was part of their life. But what has an anonymous feltmaker in the Middle East, who in his yurt lies down on his felt carpet and looks up through the round roof-hole, in common with a young felt fashion designer with a big name? The sky is their limit.

Künstler-Biographien

Biographies of the Artists

Ulrich Beckert
»the walking house«
München (D)

Geboren 1966. Studium an der Akademie der Bildenden Künste Nürnberg, Architektur und Design bei Prof. Uwe Fischer. Seit 1997 zusammen mit Georg ▷ Soanca-Pollak eigenes Designstudio in Nürnberg und München. Seit Januar 2000 vertreiben beide ihre Möbel, Leuchten und Accessoires unter dem Label »the walking house«.
Born in 1966. Studied architecture and design under Professor Uwe Fischer at the Nuremberg Akademie der Bildenden Künste. In 1987 together with ▷ Georg Soanca-Pollak he founded a design studio in Nuremberg and Munich. Since January 2000 they have been marketing their furniture, lighting fixtures and accessories under the label 'the walking house'.

Heather Belcher
London (GB)

Geboren 1960. Studium im Fachbereich Textil des Goldsmiths' College London 1980–83 (B.A.) und 1992–95 (M.A.). Mitglied des Crafts Council Selected Index of Masters.
Born in 1960. Studied Textiles at Goldsmiths' College London 1980–83 (B.A.) and 1992–95 (M.A.). Member of the Crafts Council Selected Index of Masters.

Denise Bettelyoun
Otzberg (D)

Geboren 1967 in Heidelberg. Studium der Kunstgeschichte, Vor- und Frühgeschichte und Anglistik in Kiel und Heidelberg, 1991–93. Seit 1993 Studium im Bereich der Freien Gestaltung an der Hochschule für Gestaltung Offenbach. 1996/97 einjähriges Materialstipendium der Johannes-Mosbach-Stiftung der Stadt Offenbach, 1997 Charlotte-Prinz-Stipendium der Stadt Darmstadt.
Born in Heidelberg in 1967. Studied art history, prehistory, early history and English in Kiel and Heidelberg, 1991–93. From 1993 studied art and design at the Hochschule für Gestaltung Offenbach. 1996/97 one-year bursary from the Johannes-Mosbach-Stiftung of the city of Offenbach, 1997 Charlotte-Prinz-Stipendium bursary from the city of Darmstadt.

Lies Bielowski
Innsbruck (A)

Geboren 1958. Seit 1988 als freischaffende Künstlerin tätig.
Born in 1958. Has worked as a freelance artist since 1988.

Christine Birkle
»HUT up«
Berlin (D)

Geboren 1962. Studium Industrie- und Modedesign an der Hochschule der Künste (HdK) Berlin 1983–93. Eignes Atelier »HUT up« seit 1993. Anfangs hauptsächlich auf Hüte ausgerichtet, wurde das Angebot bald um weitere Accessoires erweitert. Die erste Kleiderkollektion wurde 1996 in Paris vorgestellt. Mittlerweile werden ihre Kollektionen weltweit, vor allem in Japan, Großbritannien, Italien, Frankreich und den USA in exklusiven Läden und Warenhäusern vertrieben.
Born in 1962. Studied industrial and fashion design at HdK (University of Fine Arts) Berlin 1983–93. After finishing her studies in 1993 she directly launched her label HUT up. In the beginning her main focus was on millinery, but soon she started creating other accessories. The first collection of clothes was presented in 1996 in Paris. Meanwhile her products are available in exclusive shops and department stores all over the world, especially in Japan, USA, Great Britain, Italy and France.

Célio Braga
Amsterdam (NL)

Geboren 1965 in Guimarania/Brasilien. Studium der Grafik und Malerei an der Boston Museum School, Boston/U.S.A., 1988–90,

Schmuckgestaltung an der Gerrit Rietveld Akademie Amsterdam 1996–2000.
Born in 1965 in Guimarania/Brazil. Studied graphic design and painting at the Boston Museum School, Boston, Mass., 1988–90, jewellery design at the Gerrit Rietveld Academy in Amsterdam 1996–2000.

Charlotte Buch
Hvalsø (DK)

Geboren 1956. Ausbildung als Kürschnerin, 1978. Hirtin, 1979. Studium an der Konsthandværkerskolen Kopenhagen, Textildesign, 1986. Lehrtätigkeit in der Erwachsenenbildung seit 1998.
Born in 1956. Trained as a furrier, 1978. Shepherdess, 1979. Studied at Konsthandværkerskolen Copenhagen, Textile Design, 1986. Instructor in adult education since 1998.

Claudia Clavuot-Merz
Chur (CH)

Geboren 1962 in Chur. Lehrerseminar Chur, Abschluß 1983. Lehrerin an der Primarschule in Domat/Ems 1983–87. Ausbildung als Werklehrerin an der Schule für Gestaltung Zürich 1987–89. Werklehrerin an der Bündner Frauenschule 1989–94. Seit 1989 eigenes Atelier in Chur.
Born in Chur in 1962. Teachers' training in Chur, graduated in 1983. Taught primary school in Domat,

Ems 1983–87. Trained as a crafts teacher at the Schule für Gestaltung, Zurich, 1987–89. Instructor at Bündner Frauenschule 1989–94. Has had her own studio in Chur since 1989.

Carlo Contin
Limbiate/MI (I)

Geboren 1967 in Limbiate (MI). Nach praktischer Tätigkeit im Möbelhandel seit 1998 eigenes Studio für Design- und Inneneinrichtung. Entwürfe für den Museumsshop des Museum of Modern Art New York sowie für die Firmen Ravanini & Castoldi und Meander (Niederlande).
Born in 1967 in Limbiate, Italy. After working in furniture stores, he set up his own design and interior decorating studio in 1998. Has designed for the Museum Shop at the New York Museum of Modern Art as well as Ravanini & Castoldi and Meander (the Netherlands).

Bernadette Ehmanns
Düsseldorf (D)

Geboren 1953 in Welschen-Ennest/Sauerland. Pädagogische Arbeit mit Kindern und Erwachsenen 1970–80. Studium an der Fachhochschule Krefeld, Produktdesign, Schwerpunkt Textildesign 1980–88. Seit 1988 selbständige Tätigkeit, u.a. im Designstudio Hoff, Krefeld, beschäftigt mit Konzeption und Entwurf im Druck- und Webbereich für Heimtextilien. Seit

1999 Produktdesignerin bei ▷ HEY-Sign, maßgeblich verantwortlich für die Kollektionen Wohnaccessoires und Modeaccessoires aus Filz.
Born in 1953 in Welschen-Ennest, Sauerland. Taught children and adults 1970–80. Studied at Fachhochschule Krefeld: product design, specialisation textile design 1980–88. Since 1988 freelance, for Designstudio Hoff, Krefeld, and others; worked on conception and design in printing and weaving domestic textiles. Since 1999 product designer for ▷ HEY-Sign, responsible for the Felt Living and Fashion Accessories Collection.

Inge Evers
Haarlem (NL)

Nach dem Besuch des Gymnasiums Studium an Instituut voor Perswetenschap 1960–62: Zeitungsgeschichte, Praktischer Journalismus, Presserecht. Daneben praktische Mitarbeit beim »Nieuwe Provinciale Groninger en Drentsche Courant«. Seit 1974 künstlerisch vor allem mit Textilien und Papier tätig; Zusammenarbeit mit Künstlern in verschiedenen europäischen Ländern, den USA und Japan. Gastdozentin bei zahlreichen internationalen Konferenzen und Symposien.
After attending secondary school, studied at the Instituut voor Perswetenschap 1960–62: history of journalism, practical journalism, press law; also worked at 'Nieuwe Provinciale Groninger en Drentsche

Courant'. From 1974 worked as an artist, mainly in textiles and paper; collaboration with artists in Europe, the USA and Japan. Guest lecturer at numerous international conferences and symposia.

Ralph Feiner
Chur (CH)

Geboren 1961. Lehrerseminar Chur, Abschluß 1981. Sprachschule in Portugal 1982, danach längerer Aufenthalt auf den Azoren und in Brasilien. Besuch von Workshops für Fotografie an der Schule für Gestaltung Zürich 1983. Ausstellungstechniker am Naturmuseum Chur 1984–89. Seit 1989 nach Aufenthalten in China und im Jemen als Landschafts- und Architekturfotograf für kulturelle Institutionen und Fachzeitschriften tätig. Seit 1995 auch Gestaltung von Möbeln, teilweise in Zusammenarbeit mit ▷ Claudia Clavuot-Merz.
Born in 1961. Graduated from teachers' training seminar in Chur in 1981. Language school in Portugal 1982, long stays on the Azores and in Brazil. Attended photography workshops at the Schule für Gestaltung, Zurich, in 1983. Exhibition technologist at Naturmuseum Chur 1984–89. After stays in China and Yemen, worked as a landscape and architecture photographer for cultural institutions and specialist publications from 1989. Since 1995 has designed furniture, also in collaboration with ▷ Claudia Clavuot-Merz.

India Flint
Mount Pleasant (AUS)

Geboren 1958 in Melbourne. Besitzt die lettische und australische Staatsangehörigkeit. Studium an der University of South Australia. Commonwealth Stipendium 1973. S.A.C.A.T. Arts Access Grant, University of South Australia Travel Grant, Allport Writing Award, Helpmann Academy Grant 1999. Abschluß des Studiums mit Master of Art 2000.
Born in Melbourne 1958. Is a dual Latvian and Australian national. Studied at the University of South Australia. Commonwealth Scholarship 1973. S.A.C.A.T. Arts Access Grant, University of South Australia Travel Grant, Allport Writing Award, Helpmann Academy Grant 1999. Completing an MA degree in 2000.

Lene Frantzen
Kolding (DK)

Geboren 1961 in Singapore. Studium an der Konsthandvaerkerskolen Kopenhagen 1980–84. Stipendien der Johannes Krøiers Fondation 1985, Denmark Nationalbanks Fondation 1989, 1994, 1999, Knud Højgaards Fondation 1994, L.F. Fogths Fondation 1999, Center for Danisk Art 1999. Lehrt seit 1985 an der Art School in Kolding.
Born in 1961 in Singapore. Studied at Konsthandvaerkerskolen Kopenhagen 1980–84. Grants from the Johannes Krøiers Fondation 1985, Denmark Nationalbanks Fondation

1989, 1994, 1999, Knud Højgaards Fondation 1994, L.F. Fogths Fondation 1999, Center for Danisk Art 1999. Has taught at Art School in Kolding since 1985.

Cristina Fröhlich
Dürnten (CH)

Geboren 1965. Ausbildung zur Werklehrerin an der Hochschule für Gestaltung Zürich 1984–89. Auszeichnung mit dem eidgenössischen Preis für Gestaltung 1992.
Born in 1965. Trained as a crafts teacher at the Hochschule für Gestaltung Zurich 1984–89. Award: Eidgenössischer Preis für Gestaltung 1992.

Bernhard Früh
Erfurt (D)

Geboren 1949 in Erfurt. Berufsabschluß als Feinmechaniker 1969. Abitur 1973. Berufsabschluß als Werbegestalter 1979. Seit Anfang der 80er Jahre Autodidakt als Schmuck- und Metallgestalter. Seit 1986 freiberuflich tätig.
Born in 1949 in Erfurt. Trained as a precision engineer, certificate in 1969. Examinations qualifying for university entrance in 1973. Trained as an advertising designer, certificate in 1979. Since the early 1980s a self-taught jewellery and metal designer. Has worked as a freelance since 1986.

Verena Gloor
Zürich (CH)

**Geboren 1927. Lehrabschluß
im Textilbereich 1945, weitere
Tätigkeit im Textilbereich
1945–51. Maturität 1953. Schule
für Gestaltung Zürich, Vorkurs
1953/54, Textilklasse 1954–57.
Lehrtätigkeit an der Gewerbeschule Zürich 1955–59, am
Werkseminar der Schule für
Gestaltung Zürich 1958–66.
Klassenlehrerin der Fachklasse
Textilgestaltung für Werklehrer
an der Schule für Gestaltung
1966–87, seit 1980 auch Filzherstellung. Seit 1976 zahlreiche
Reisen zu Filzworkshops und
-symposien.**
Born in 1927. Finished apprenticeship in textiles in 1945, worked in
textiles 1945–51. University
entrance qualification examinations
in 1953. Schule für Gestaltung
Zurich, preliminary course 1953/54,
textile class 1954–57. Taught at
Gewerbeschule Zurich 1955–59, at
the Werkseminar der Schule für
Gestaltung Zurich 1958–66. Form
teacher for the specialist class in
textile design for crafts teachers at
the Schule für Gestaltung 1966–87,
since 1980 also felting. Since 1976
has travelled to many felt workshops and symposia.

Hermine Gold
Pinzberg (D)

**Geboren 1945 in
Oberrodach/Kronach. Studium
an der Pädagogischen Hochschule Würzburg 1964–67. Seit**
**1990 Beschäftigung mit Mate-
rialarbeiten und Malerei
als Autodidaktin.**
Born in 1945 in Oberrodach,
Kronach. Studied at the Education
Department of the University of
Würzburg 1964–67. Since 1990 has
worked with materials and painting
as a self-taught artist.

Heidi Greb
Ottobrunn (D)

**Geboren 1968 in München.
Abitur 1988. Damenschneider-
lehre in München 1988–91.
Studium Modedesign an der
Fakultät Angewandte Kunst der
Westsächsischen Hochschule
Zwickau in Schneeberg/Sach-
sen 1993–98, Diplom 1998.
Seitdem freiberuflich selbstän-
dig mit eigenem Atelier »Filz –
Kleider – Kunst«. 4. Preis beim
3. Europäischen Wettbewerb
für Textil und Modedesign,
Apolda/Thüringen 1999. Ehren-
preis der Danner-Stiftung
München 1999. Artist in Residence, Koli, Finnland,
1999/2000.**
Born in Munich in 1968. University
entrance qualification examinations
in 1988. Apprenticeship as a
dressmaker in Munich 1988–91.
Studied fashion design at the
Department of Applied Arts of the
Westsächsische Hochschule
Zwickau in Schneeberg/Saxony,
1993–98. Graduated in 1998. Has
worked since then as a freelance
artist with her own studio. Won
4th prize at the 3rd European Competition for Textile and Fashion
Design, Apolda/Thuringia. Hon-
orary award of the Danner-Stiftung
Munich 1999. Artist in Residence,
Koli, Finland, 1999/2000.

Ulrike Hein
Köln (D)

**Geboren in Neuenbürg 1960.
Gesellenprüfung als Gärtnerin
1982. Studium an der Staat-
lichen Akademie der Bildenden
Künste Stuttgart bei Jürgen
Brodwolf. Akademiepreis 1988.
Förderpreis der Bildenden Kunst
des Bundesministeriums für
Bildung und Wissenschaft 1989.
Atelier-Stipendium des Landes
Baden-Württemberg 1991–93.
Stipendium der Kunststiftung
Baden-Württemberg 1992.
Friedrich-Vordemberge-Stipen-
dium der Stadt Köln 1993.
Stipendiatin an der Cité Interna-
tionale des Arts, Paris 1995.
Seit 1995 Mitglied im Bund Bil-
dender Künstler (BBK).**
Born in 1960 in Neuenbürg.
Finished an apprenticeship as
a gardener in 1982. Studied under
Jürgen Brodwolf at the Staatliche
Akademie der Bildenden Künste
in Stuttgart. Bursary of Bildende
Kunst from the Federal Mininstry
for Education and Science 1989.
Atelier grant from Baden-Württem-
berg 1991–93. Bursary from the
Kunststiftung Baden-Württemberg
1992. Friedrich-Vordemberge-
Stipendium bursary from the city
of Cologne 1993. Atelier grant from
Cité Internationale des Arts, Paris
1995. Since 1995 member of the
Bund Bildender Künstler (BBK).

HEY-SIGN
Matthias Hey
Düsseldorf (D)

▷ **Bernadette Ehmanns**

Käthi Hoppler-Dinkel
Ostermundigen (CH)

**Geboren 1950. Studium an der
Kunstgewerbeschule Zürich
1966–71. Kennenlernen der Filztechnik 1982. Seit 1984 eigenes
Filzatelier in Ostermundigen
bei Bern.**
Born in 1950. Studied at Kunstgewerbeschule Zurich 1966–71.
Became acquainted with felting in
1982. Since 1984 has had a felt
studio of her own in Ostermundigen near Bern.

Elina Huotari
Helsinki (FIN)

**Geboren 1970. Studium an der
University of Art and Design
Helsinki, Fakultät für Mode- und
Textildesign, Abschluß mit
Masters-Diplom 1999. Assistant
Designer bei Ratti spa, Como
1996, dort zusammen mit ▷ Tuttu
Sillanpää 1. Preis beim Ratti
European Textile Design Work-
shop Competition. Seit 1999 als
Designerin für die von ▷ Kirssika
Savonen geleitete Firma Verso
Design Oy tätig.**
Born in 1970. Studied at the Univer-
sity of Art and Design Helsinki,
Department of Fashion and Textile
Design, took her Master of Arts
degree in 1999. Assistant Designer

at Ratti spa, Como 1996, where she
won the 1st prize of the Ratti European Textile Design Workshop
Competition together with ▷ Tuttu
Sillanpää. Since 1999 she has
worked as a designer for ▷ Kirssika
Savonen's firm Verso Design Oy.

Hillevi Huse
Oslo (N)

**Geboren 1955. Studium an der
Königlichen Kunstakademie
Oslo 1983–87. Mitglied der nor-
wegischen Künstlervereinigung.**
Born in 1955. Studied at the Royal
Art Academy in Oslo 1983–87.
Member of the Norwegian Artists'
Association.

May Jacobsen Hvistendahl
Furnes (N)

**Geboren 1949 in Raufoss,
Norwegen. Ausbildung als Jour-
nalistin. Seit 1985 professionelle
Filzerin. Gründete 1995 die
Gruppe »Norske Filtmakere«,
der sie drei Jahre lang vorstand.
Einladung zu Lehrveranstaltungen anlässlich des »Europeisk Kulturby« in Bergen und
zum Australian Forum for Textile
Arts 2000.**
Born in 1949 in Raufoss, Norway.
Trained as a journalist. Since 1985
professional felter. Founded the
'Norske Filtmakere' group in 1995,
which she was head of for three
years. Invited to teach in the framework of the 'Europeisk Kulturby'
in Bergen and at the Australian
Forum for Textile Arts 2000.

Jorie Johnson
Kyoto (Japan)

**Geboren 1954 in Boston, USA.
Studium an der Rhode Island
School of Design 1976–77, am
Pohjois-Savo Polytechnic in
Kuopio, Finnland, 1977–78, Tai-
deteollinenkorkeakoulu (Hoch-
schule für Kunst und Design),
Helsinki 1983–84. Gründung des
Design-Studios Joi Rae™ Textiles
in Boston 1979, verlegt nach
Kyoto 1988. Lehraufträge am
Kyoto College of Art 1997–99, an
der Kyoto University of Art and
Design seit 2000.**
Born in 1954 in Boston, USA. Studied at Rhode Island School of
Design 1976–77, Pohjois-Savo Polytechnic in Kuopio, Finland, 1977–78,
Taideteollinenkorkeakoulu (University of Art and Design), Helsinki
1983–84. Joi Rae™ Textile Design-
Studio established in Boston 1979,
moved to Kyoto 1988. Part-time
lecturer in the Department of Textile Design Kyoto College of Art
1997–99, at Kyoto University of Art
and Design since 2000.

Anna-Kaarina Kalliokoski
Isokyrö (FIN)

**Geboren 1962. Studium am
Institut für Kunsthandwerk und
Design in Kuopio (Kuopion
käsi- ja taideteollisuusoppiatos)
1987, Master-Studiengang an
der Hochschule für Angewandte
Kunst Helsinki (Taideteollinen
korkeakoulu) seit 1990.
De Montfort University 1993.**
Born in 1962. Studied at the Insti-

tute for Crafts and Design in Kuopio in 1987, Master of Arts at College of Applied Arts Helsinki since 1990. De Montfort University 1993.

Reto Kaufmann
»Einzigart« – Plattform für
Design
Zürich (CH)

Geboren 1965. Lehre als Möbelschreiner. Kunstschule Form + Farbe, Zürich. Seit 1991 Vertrieb von Möbeln und Wohnaccessoires unter eigenem Label »Jaesooepis«. Seit 1996 »Einzigart – Plattform für Design«, Möbelgeschäft mit eigenen Entwürfen und Gästen.
Born in 1965. Served an apprenticeship as a carpenter and joiner. Kunstschule Form + Farbe, Zurich. Since 1991 has been marketing his own furniture and living accessories under the 'Jaesooepis' label. Since 1996 he has had a furniture shop, 'Einzigart – Plattform für Design', where he sells furniture he and others design.

Ernst Köster
Meschede (D)

Geboren 1958 in Meschede. Abitur 1978. Steinmetzlehre 1979–81. Westfälisches Amt für Denkmalpflege, Restauratorenausbildung 1981–84. Kunstakademie Düsseldorf, Abt. Münster 1984–88. Seit 1988 freischaffend tätig.
Born in 1958 in Meschede. Examinations qualifying for university

entrance in 1978. Apprenticeship as a stonemason 1979–81. Trained as a restorer at Westfälisches Amt für Denkmalpflege 1981–94. Kunstakademie Düsseldorf, Div. Münster 1984–88. Freelance since 1988.

Riona Kuthe
Nürtingen (D)

Abitur 1979. Ausbildung zur Weberin und Webgestalterin 1981–83. Studium der Kunsterziehung 1983–85. Studium der Textilkunst an der Akademie der Bildenden Künste Nürnberg 1985–91, Meisterschülerin von Hans Herpich.
Examinations qualifying for university entrance in 1979. Trained as a weaver and woven textiles designer 1981–83. Trained as an art teacher 1983–85. Studied textiles at the Akademie der Bildenden Künste Nuremberg 1985–91, master class Hans Herpich.

Turid Kvalvåg
Alta (N)

Geboren in Tønsberg/Norwegen 1950. Studium der Angewandten Kunst. Etwa 20 Jahre als Lehrerin in verschiedenen Schulstufen tätig. Seit 1995 Mitglied der Norwegian Feltmakers Association.
Born in 1950 in Tønsberg, Norway. Studied applied arts. Has taught at schools at various levels for about 20 years. Since 1995 membership of the Norwegian Feltmakers' Association.

Paola Lenti
Meda/MI (I)

Geboren 1958 in Alessandria, Piemont. Studium am Design Polytechnicum in Mailand u. a. bei Bruno Munari und Hans Ballmer. Anfang der 80er Jahre Eröffnung eines eigenen Ateliers für Graphic Design in Meda. Erste Porzellankollektion 1994 in Mailand vorgestellt. Gründung der Firma Paola Lenti für Wohnaccessoires, in der seit 1995 auch Filzteppiche und Sitzmöbel mit Filzbezügen hergestellt werden.
Born in 1958 in Alessandria, Piemonte. Studied at the Design Polytechnic in Milan, teachers among others Bruno Munari and Hans Ballmer. Opened a graphic design studio in Meda in the early 1980s. In 1994 the first collection of biscuit porcelain was exhibited in Milan. Founded the company 'Paola Lenti', where she designs and produces accessories for the house, since 1995 also felt carpets and pouffes.

Ute Lindner
Berlin (D)

Geboren in Arnsberg/Westfalen 1968. Studium der Kunstwissenschaft und Philosophie in Kassel 1989–95, parallel dazu ab 1990 Studium der Freien Kunst, Abschluß mit Magister Artium und Diplom 1995. Volontariat am Institute of Contemporary Arts, London. Stipendium des Deutsch-Französischen Jugendwerks und Atelier-Stipendium der Cité Internationale des Arts, Paris 1996. Kasseler Kunstpreis 1996. Umzug nach Berlin 1997. Tutorial am Goldsmiths' College, London. Mitherausgeberin des Kunstmagazins »Copyright« 1999.

Born in 1968 in Arnsberg, North Rhine-Westphalia. Studied art history and philosophy in Kassel 1989–95, from 1990 also studied art, graduated with an MA degree and a diploma in 1995. Volunteer at the Institute of Contemporary Arts, London. Bursary from the Deutsch-Französisches Jugendwerk and Atelier grant from Cité Internationale des Arts, Paris 1996. Kasseler Kunstpreis/Art Award 1996. Moved to Berlin in 1997. Tutor at Goldsmiths' College London. Co-editor of the art magazine 'Copyright' in 1999.

Heyke Manthey
Basthorst (D)

Geboren 1954 in Wankendorf/Schleswig-Holstein. Restauratorin bei Antik C. & M. Broocks, Gettorf. Studium an der Hochschule der Künste Berlin, Hauptfach Textilgestaltung 1989–94, seit 1993 auch in der Kostümbildklasse. Arbeiten mit Filz seit 1991. Freiberuflich tätig als Textildesignerin seit 1995. Eröffnung eines Textilateliers auf Gut Basthorst 1996. Zusammenarbeit mit der Fa. Gottstein, Österreich, zur industriellen Umsetzung der Filzkollektion, daneben Filz- und

Seidenstoffunikate für Privatkunden und Theaterausstattungen, wie »Kohlhaas« von Harald Müller im Ernst-Deutsch-Theater, Hamburg 1996 und »Land des Lächelns« von Franz Lehár im Forum Junges Musiktheater, Hamburg 1998.

Born in 1954 in Wankendorf, Schleswig-Holstein. Restorer for Antik C. & M. Broocks, Gettorf. Studied at Hochschule der Künste Berlin, specialisation textile design 1989–94, from 1993 also in the class for stage costume design. Has worked with felt since 1991. Freelance textile designer since 1995. Opened a textile studio at Basthorst Manor in 1996. Collaborated with the firm of Gottstein, Austria, on industrial uses of felt collection, also on one-of-a-kind felts and silks for private clients and stage productions, among them Harald Müller's 'Kohlhaas' at the Ernst-Deutsch-Theater in Hamburg 1996 and Franz Lehar's 'Land des Lächelns' at Forum Junges Musiktheater in Hamburg 1998.

Claudia Merx
Stolberg (D)

Geboren 1957 in Mönchengladbach. Studium an der Fachhochschule Niederrhein, Mönchengladbach, Fachbereich Textilgestaltung, 1977–83. Selbständig tätig als Textildesignerin, Mitglied der Gruppe »Kunstwerk« 1983–88. Atelier für Textilkunst seit 1997.

Born in 1957 in Mönchengladbach. Studied at Fachhochschule Nieder-

rhein, Mönchengladbach, specialising in textile design, 1977–83. Worked as a freelance textile designer, member of the 'Kunstwerk' Group 1983–88. Has had her own studio for textile art since 1997.

Mari Nagy
Kecskemét (H)

Geboren 1948 in Budapest, verheiratet mit ▷ István Vidák. Beide erhielten ihre Ausbildung am Lehrerinstitut in Kecskemét. Gemeinsam erforschten sie seit 1974 alte Handwerkstechniken und veranstalteten am Szórakaténusz Spielzeugmuseum Kecskemét von 1982–98 praktische Kurse im Korbflechten und Spielzeugschnitzen. Seitdem sind beide als freischaffende Künstler tätig.

Born in Budapest in 1948, married to ▷ István Vidák. Both trained at Kecskemét Teachers' Training Institute. Since 1974 they have been studying traditional crafts techniques and taught practical courses in basket weaving and toy carving at Szórakaténusz Toy Museum in Kecskemét 1982–98. Since then both have worked as freelance artists.

Kay Eppi Nölke
Pforzheim (D)

Geboren 1960 in Bielefeld. Goldschmiedelehre 1980–83, anschließend als Goldschmied tätig. Zeichenakademie Hanau

1989–91: Abschluß als staatl. gepr. Gestalter, Meisterprüfung. Studium an der FH Pforzheim, Hochschule für Gestaltung 1993–98, Abschluß mit Diplom. Seitdem freischaffend tätig.
Born in 1960 in Bielefeld. Apprenticeship as a goldsmith 1980–83, then worked as a goldsmith. Zeichenakademie Hanau 1989–91: graduated as a designer with state examinations, master certificate. Studied at FH Pforzheim, Hochschule für Gestaltung 1993–98. Since graduation has worked as a freelancer.

Anja Oehler
Washington, D.C. (USA)

Geboren 1968 in Stuttgart. Abitur 1988. Freie Kunstschule Stuttgart 1989. Studium der Innenarchitektur an der Staatlichen Akademie der Bildenden Künste in München 1990–96. Architekturbüro Nalbach, Berlin, Mitarbeit am Projekt »art'otel Ermeierhaus« Berlin 1996–97. Professional Diploma in Architecture, University of North London 1997–99. Architekturbüro Reena Racki Associates, Washington, D.C., seit 2000.
Born in 1968 in Stuttgart. Examinations qualifying for university entrance in 1988. Freie Kunstschule Stuttgart 1989. Studied interior decoration at the Staatliche Akademie der Bildenden Künste in Munich 1990–96. Worked for the architectural firm of Nalbach, Berlin, collaborated on the 'art'otel Ermeierhaus' project in Berlin

1996–97. Studied at the University of North London 1997–99, graduating with a professional diploma in architecture. Has worked for Reena Racki Associates, architects, in Washington, D.C., since 2000.

Gunilla Paetau Sjöberg
Bålsta (S)

Geboren 1952 in Åbo, Finnland. Studium am Handicraft Institute in Borgå, Finnland, 1960–62, an der Schule für Textiles Werken in Uppsala, Schweden, 1978–79, Pädagogik, Ethnographie und Kunstgeschichte an der Universität Uppsala, 1982–89. Seit 1973 Lehrtätigkeit an verschiedenen Hochschulen in Schweden.
Born in 1952 in Åbo, Finland. Studied at the Handicraft Institute in Borgå, Finland, 1960–62, at the School for Textiles in Uppsala, Sweden, 1978–79, education, ethnology and art history at Uppsala University, 1982–89. Since 1973 she has taught at various Swedish colleges and universities.

Silja Puranen
Järvenpää (FIN)

Geboren 1961 in Elimäki/Finnland. Studium am Institut für Kunsthandwerk und Design in Kuopio (Kuopion käsi- ja taideteollisuusoppiaitos), Textildesign, 1987.
Born in 1961 in Elimäki, Finland. Studied at the Institute of Crafts and Design in Kuopio, textile design, 1987.

Kirsikka Savonen
Verso Design Oy
Helsinki (FIN)

Geboren 1943 in Lahti. Studium der Innenarchitektur am Institute of Industrial Arts Helsinki, Diplom 1969. Designerin für Asko Oy 1971–74. Seit 1974 selbständig tätig als Möbeldesignerin und Produktentwicklerin für verschiedene finnische Firmen. Seit 1997 Managing Director von Verso Design Oy. Vorstandsmitglied des Verbands Finnischer Innenarchitekten 1983–89, Vorsitzende 1989. Vorstandsmitglied des Finnischen Designerverbands ORNAMO1988–91. Eeva und Maija Taimi- Stipendium des Finnischen Designerverbands 1996. Für einen aus Holz- und Leinenstreifen gefertigten Teppich wurde sie zusammen mit ▷ Tuttu Sillanpää u. a. mit dem Internationalen Desingpreis Baden-Württemberg 1999 ausgezeichnet.
Born in 1943 in Lahti. Studied at the Institute of Industrial Arts, Helsinki, diploma as Interior Architect 1969. Designer for Asko Oy 1971–74. Independent entrepreneur since 1974. Designed furniture and was member of product developing groups for several Finnish companies. Managing director of Verso Oy since 1997. Member of the board of the Finnish Association of Interior Architects SIO 1983–89, chairwoman 1989. Member of the board of the Finnish Association of Designers ORNAMO 1988–91. Eeva and Maija Taimi scholarship of

ORNAMO 1996. Won several prizes for a wooden strip / linen carpet together with ▷ Tuttu Sillanpää, among others the Baden-Württemberg International Design Award 1999.

Jeanette Sendler
Edinburgh (GB)

Geboren 1965 in Berlin. Ausbildung als Damenmaßschneiderin, danach vier Jahre in der Kostümwerkstatt der Komischen Oper Berlin. Seit 1991 in Schottland tätig. Studium am Edinburgh College of Art (Kostümbildnerei), Master-Abschluß 1997. Während des Studiums Kostümassistentin für die Scottish Opera, English National Opera und Australian Opera (Sydney). Seit 1995 Arbeit an interaktiven Kostümen, woraus die erste Performance »Light Enlightens« entstand. 1997 Gründung der Gruppe »Metacorpus« zusammen mit Anna Cocciadiferro und Chloë Dear. Metacorpus repräsentiert Kostümkunst als eine eigenständige Kunstform.
Born in Berlin in 1965. Her initial training as a ladies' dressmaker was followed by four years as a costume maker for the Comic Opera, Berlin. Since 1991 Jeanette has been based in Scotland. She studied Costume Design at Edinburgh College of Art and completed her MA in 1997. Whilst studying, Jeanette worked as a freelance for the Scottish Opera, the English National Opera

and the Australian Opera (Sydney). In 1995 Jeanette first began working on interactive costumes, which led to the performances of 'Light Enlightens'. In 1997 she founded Metacorpus together with Anna Cocciadiferro and Chloë Dear, a company which concentrates on increasing awareness and appreciation of costume art.

Tuttu Sillanpää
Helsinki (FIN)

Geboren 1967. Besuch der Helsinki Senior Secondary School of Visual Arts, Abschluß 1986. Seit 1988 Studium an der University of Art and Design Helsinki (UIAH), Fakultät für Mode- und Textildesign sowie an der Fakultät für Kunsterziehung; Master of Arts 1995. Seit 1993 Lehrerin für Kunst und Textildesign an der Helsinki Senior Secondary School of Visual Arts. Assistant Designer bei Ratti spa, Como 1996, dort zusammen mit ▷ Elina Huotari. 1. Preis beim Ratti European Textile Design Workshop Competition. Seit 1999 als Designerin für die von ▷ Kirssika Savonen geleitete Firma Verso Design Oy tätig. Internationaler Designpreis Baden-Württemberg 1999.
Born in 1967. Studied at Helsinki Senior Secondary School of Visual Arts, High School Graduate 1986. Studied at the University of Art and Design Helsinki (UIAH), Department of Fashion and Textile Design, since 1988, took her Master of Arts degree at the Department of

Art Education in 1995. Teacher of art and textile design at the Helsinki Senior Secondary School of Visual Arts, since 1993. Assistant Designer at Ratti spa, Como 1996, where she won the 1st prize of the Ratti European Textile Design Workshop Competition together with ▷ Elina Huotari. Since 1999 she has worked as a designer for ▷ Kirssika Savonen's firm Verso Design Oy. International Design Prize Baden-Württemberg 1999.

Georg Soanca-Pollak
»the walking house«
München (D)

Geboren 1967. Studium an der Akademie der Bildenden Künste Nürnberg, Architektur und Design bei Prof. Uwe Fischer. Seit 1997 zusammen mit ▷ Ulrich Beckert eigenes Designstudio in Nürnberg und München. Seit Januar 2000 vertreiben beide ihre Möbel, Leuchten und Accessoires unter dem Label »the walking house«.
Born in 1967. Studied architecture and design under Professor Uwe Fischer at the Nuremberg Akademie der Bildenden Künste. In 1987 together with ▷ Ulrich Beckert he founded a design studio in Nuremberg and Munich. Since January 2000 they have been marketing their furniture, lighting fixtures and accessories under the label 'the walking house'.

Annette Stoll
Weinböhla (D)

**Geboren 1978 in Dresden. Nach
dem Abitur Studium der Textilkunst an der Fakultät Angewandte Kunst der Westsächsischen Hochschule Zwickau in
Schneeberg 1996–2000. Aus-
tauschsemester in Joensuu/
Finnland 1998, Erlernen der Filztechnik.**
Born in 1978 in Dresden. After
qualifying for university entrance,
studied textile design at the
Department of Applied Arts of the
Westsächsische Hochschule
Zwickau in Schneeberg 1996–2000.
Spent a semester as an exchange
student in Joensuu, Finland, in
1998, where she learnt felting.

Ilkka Suppanen
Helsinki (FIN)

**Geboren 1968 in Kotka/Finn-
land. Studium der Architektur an
der Technischen Hochschule
Helsinki 1988, Design-Studium
an der University of Art and
Design Helsinki 1989–92, und
an der Gerrit-Rietveld-Akademie
Amsterdam 1992. Seit 1995
eigenes Studio in Helsinki als
frei schaffender Designer u. a.
für Cappellini, Proventus und
Saab. Mitbegründer der Design-
Cooperative Snowcrash 1996.
Lehrauftrag an der University of
Art and Design Helsinki
1995–2000. 1998 von Ettore
Sottsass als einer der vier wich-
tigsten jungen Designer für den
Dedalus-Preis nominiert.**

Born in 1968 in Kotka, Finland.
Studied architecture at the Technical University of Helsinki, 1988,
design at the University of Art and
Design Helsinki 1989 – 92, and
at the Gerrit Rietveld Academy
Amsterdam 1992. He has worked
as a freelance designer among
others for Cappellini, Proventus,
Saab. In 1996 with his colleagues,
he founded the Snowcrash design
co-operative. He taught design
at the University of Art and Design
Helsinki 1995–2000. 1998 he was
nominated for Dedalus prize for
young European Designers by
Ettore Sottsass as one of the four
most important young designers
under 40.

Ines Tartler
Amsterdam (NL)

**Geboren 1969 in Esslingen am
Neckar. Ausbildung zur Augen-
optikerin 1988–91. Berufskolleg
für Formgebung Schmuck und
Gerät, Schwäbisch Gmünd,
1992–95. Fachhochschule für
Gestaltung Pforzheim, Fachbereich Schmuck und Gerät
1996–98. Seit 1998 Gerrit Rietveld Akademie Amsterdam:
Schmuck**
Born in 1969 in Esslingen on the
Neckar. Trained as an optician
1988–91. Studied at Berufskolleg für
Formgebung Schmuck und Gerät,
Schwäbisch Gmünd, 1992–95 and
at Fachhochschule für Gestaltung
Pforzheim, Department of Jewellery
and Metalwork 1996–98. Since
1998 at the Gerrit Rietveld Academy in Amsterdam: jewellery.

Ingrid Thallinger
Linz (A)

**Geboren 1962 in Gmunden. Aus-
bildung: Kolleg für Mode und
Bekleidungstechnik in Ebensee
1989–91. Seit 1997 Hochschule
der Künste Linz, Meisterklasse
Textil. Sommerakademie Salzburg 1999.**
Born in 1962 in Gmunden. Training:
Kolleg für Mode und Bekleidungstechnik in Ebensee 1989–91. Since
1997 Hochschule der Künste Linz,
master class in textiles. Salzburg
Summer Academy 1999.

Katharina Thomas
Icking im Isartal (D)

**Geboren 1952 in Leipzig.
Schneiderlehre. Modeschule.
Studium an der Folkwangschu-
le/Universität Essen. Abschluß
als Diplom-Textildesignerin
1980. Seit 1982 eigenes Atelier
in Icking bei München. Lehrauftrag an der Fachoberschule für
Gestaltung München 1988–90
und an der Universität Köln
1998. 1999 Gründung der Firma
Filzkontor®. Studienaufenthalte
in der Mongolei, Türkei, Turkmenistan, Iran, den USA und
Skandinavien.**
Born in 1952 in Leipzig. Apprenticeship as a sempstress. Fashion
design school. Studied at Folk-
wangschule/Essen University.
Diploma in textile design in 1980.
Since 1982 has had her own studio
in Icking near Munich. Instructor at
the Fachoberschule für Gestaltung
in Munich 1988–90 and at Cologne

University in 1998. Founded Filz-kontor® in 1999. Travels to Mongolia, Turkey, Turkmenistan, Iran, the USA and Scandinavia.

Maisa Tikkanen
Kulennoinen (FIN)

Geboren 1952 in Oulujoki. Ausbildung: Savonlinna Senior Secondary School of Art 1968–71, Institute of Industrial Arts Helsinki 1971–75. Freischaffend als Designerin für Stoffdruck 1975–79. Lehraufträge an der University of Industrial Arts Helsinki seit 1980, Filzkurse 1988–97. Eigenes Studio für Filzteppiche seit 1992, in dem seit 1993 auch Wolle gefärbt wird. Zahlreiche Workshops im In- und Ausland.
Born in 1952 in Oulujoki, Finland. Training: Savonlinna Senior Secondary School of Art 1968–71, Institute of Industrial Arts Helsinki 1971–75. Freelance designer of textile printing 1975–79. Instructor at the University of Industrial Arts Helsinki from 1980, courses in felting 1988–97. Has had her own studio for felt carpets since 1992 and has dyed wool there since 1993. Has held numerous workshops at home and abroad.

Henriette Tomasi
Frankfurt am Main (D)

Geboren 1969 in Königstein im Taunus. Goldschmiedelehre 1989–93, Gesellenjahre 1993–95. Studium an der Zei-
chenakademie Hanau 1995–97. **Seit 1997 freischaffend im eigenen Atelier. Mitglied im Bundesverband Kunsthandwerk.**
Born in 1969 in Königstein in the Taunus. Apprenticeship as a goldsmith 1989–93, journeyman 1993–95. Studied at Zeichenakademie Hanau 1995–97. Since 1997 freelance with a studio of her own. Member of Bundesverband Kunsthandwerk.

Ingeborg Verres
Karlsruhe (D)

Geboren 1942 in Bonn. Studium der Glasgestaltung, Restauration, Malerei 1961–64. Afrikareise 1965/66. Schule für Gestaltung Siegen, Kunsttherapie, Weben 1989–92. Hochschule Bonn Bildhauerei, 1992–96. Kunstakademie Basel 1997. Filzkurs bei Istvan Vidák 1995.
Born in 1942 in Bonn. Studied glass design, trained as a restorer, studied painting 1961–64. In Africa 1965/66. Schule für Gestaltung Siegen, art therapy, weaving 1989–92. Hochschule Bonn, sculpture, 1992–96. Kunstakademie Basel 1997. Felt course under István Vidák 1995.

István Vidák
Kecskemét (H)

Geboren 1947 in Budapest; verheiratet mit ▷ Mari Nagy. Beide erhielten ihre Ausbildung am Lehrerinstitut in Kecskemét. Gemeinsam erforschten sie seit
1974 **alte Handwerkstechniken und veranstalteten am Szórakaténusz Spielzeugmuseum Kecskemét von 1982–98 praktische Kurse im Korbflechten und Spielzeugschnitzen. Seitdem sind beide als freischaffende Künstler tätig.**
Born in Budapest in 1947, married to ▷ Mari Nagy. Both trained at Kecskemét Teachers' Training Institute. Since 1974 they have been studying traditional crafts techniques and taught practical courses in basket weaving and toy carving at Szórakaténusz Toy Museum in Kecskemét 1982–98. Since then both have worked as freelance artists.

Karin Wagner
Basel (CH)

Geboren 1964 in Tansania, aufgewachsen in Brasilien und der Schweiz. Ausbildung zur Textil- und Werklehrerin am Pädagogischen Institut Basel 1986–90. Studium in der Fachklasse für Körper und Kleid der Schule für Gestaltung Basel 1991. Dozentin für gestalterisch tätige Berufe am Institut für Neues Lernen in Liechtenstein und an der Schule für Gestaltung Basel 1992/93. Kostümassistenz am Theater Frankfurt und Basel 1992 und 1998. Gaststudium der Schmuckgestaltung in Barcelona und Amsterdam 1993/94. Seit 1998 selbständig im Bereich Textildesign tätig.
Born in 1964 in Tanzania, grew up in Brazil and Switzerland. Trained

as a textile and crafts teacher at the Pädagogisches Institut in Basle 1986–90. Studied in the specialist class for clothing and the body at the Basle Schule für Gestaltung 1991. Instructor in professional design at the Institut für Neues Lernen in Liechtenstein and at the Basle Schule für Gestaltung 1992/93. Costume assistant at theatres in Frankfurt and Basle 1992 and 1998. Guest student in jewellery design in Barcelona and Amsterdam 1993/94. Has been a freelance textile designer since 1998.

Born in 1951 in Nagold. Studied at Staatliche Akademie der Bildenden Künste Stuttgart, specialisation in textile design, 1970–75. From 1984 taught 'textile design' at the Freie Kunsthochschule and the Fachhochschule für Kunsttherapie in Nürtingen and at Osnabrück University. Founder member of the 'Werk Raum Textil e.V.' group 1999. Since then has held workshop and study groups. Member of the Baden-Württemberg Verband Bildender Künstler.

The walking house
München (D)

▷ **Ulrich Beckert und** ▷ **Georg Soanca-Pollak**

Margarete Warth
Tübingen (D)

1951 in Nagold geboren. Studium an der Staatlichen Akademie der Bildenden Künste Stuttgart, Fachbereich Textildesign, 1970–75. Seit 1984 Lehrtätigkeit im Bereich »Textiles Gestalten« der Freien Kunsthochschule und der Fachhochschule für Kunsttherapie in Nürtingen sowie an der Universität Osnabrück. Gründungsmitglied der Gruppe »Werk Raum Textil e.V.«, 1999. Seither Leitung von Werkstatt- und Studienangeboten. Mitglied im Verband Bildender Künstler Baden-Württemberg.

Bibliographie
(Auswahl)

Bibliography
(A selection)

Ågren, Katarina, Toving, Stockholm 1976

Bettelyoun, Denise, bleat II, Frankfurt/M., o. J. (1999)

Bidder, Irmgard u. Hans Bidder, Filzteppiche. Ihre Geschichte und Eigenart, Braunschweig 1980 (Bibliothek für Kunst- und Antiquitätenfreunde, Bd. 56)

Brown, Victoria, Felt Work, 1996

Burkett, Mary Elizabeth, The Art of the Feltmaker, Kendal: Abbot Hall Art Gallery 1979

Damgaard, Annette, Filt, 1994

Donald, Kay, Creative Feltmaking, 1990

Ekert, Marianne, Lär dig tova, Västerås 1985

Evers, Inge, Feltmaking, London 1987 (Originalausgabe: Viltmaken, 1984)

Filzmacherei – eine uralte Kunst, in: Ciba-Rundschau 12, 1958, Nr. 139, S. 22–24

Fisher, Joan, Exploring Felting, 1997

Gervers, Michael / Veronika Gervers, Felt-Making Craftsmen on the Anatolian and Iranian Plateau, in: Textile Museum Journal, Vol. IV, Part 1, Washington D.C., 1974

Gordon, Berverly, Feltmaking, New York 1980

Hvistendahl, May Jacobsen, Håndlaget filt – Ideer og teknikker for toving av ull, Oslo 2000

Johnson, Jorie, Feltmaking – Woolmagic, 1999

Krag Hansen, Birgitte, Filt i Form, Kopenhagen 1992 (engl. 1999)

Lindner Ute, Belichtungszeiten, Darmstadt, Köln 1997

Morris, Robert, New Felt Pieces and Drawings, Hannover 1997

Nagy, Mari u. István Vidák, Filzpalette, 1999

Nagy, Mari u. István Vidák, Filzspielzeug, 3. Aufl. 1998

Nielsen, Lene, Mosekonens Filtebog, Holte (DK) 1986

Paetau Sjöberg, Gunilla, Filzen. Alte Tradition – modernes Handwerk, Bern, Stuttgart, Wien, 3. Aufl. 2000 (Originalausgabe: Tova, 1994; engl. Ausgabe: Felt. New Directions for an Ancient Craft)

Smith, Sheila / Freda Walker, Feltmaking. The Whys and Wherefores, Thrisk 1995

Smådahl, Kirsten Julie, Filting av ull, 1989

Spark, Pat, Fundamentals of Feltmaking, Washington 1989

Spark, Pat, Scandinavian Style in Feltmaking, 1992

Szeemann, Harald (Hg.), Beuysnobiscum, Neuausg, Amsterdam, Dresden 1997

Tikkanen, Maisa, Huopateoksia. Works in Felt, Helsinki 2000

Turnau, Irena, Hand-felting in Europe and Asia. From the Middle Ages to the 20th Century, Warschau: Polish Academy of Science; Institute of Archeology and Ethnology 1997

Unterlass, Ulrike: Filzen. Faszination eines alten Handwerks: Techniken & Ideen, Aarau 1996

Wolk-Gerche, Angelika, Filzen: uralte Technik, neu belebt, Peiting 1995

Zang, Ulrike, Filz, in: Reallexikon zur deutschen Kunstgeschichte, Bd. 7, München 1987, Sp. 1185 – 1196

Zeitschriften:
Surface Desing Journal, USA (Sonderheft Filz 1994)
Textilforum, Deutschland, 1982

Bildnachweis
**Alle Fotos von den Künstlern,
mit Ausnahme:**

Photo Credits
All pictures are from the artists,
except:

**Archiv Sammlung Froehlich,
Stuttgart 39
Aro, Leena 227
Artner, Norbert 85
Breig, Mimi 141
Bstieler, Markus 189, 190, 191
Degen, Thomas 221, 223
Farkas, Otto 55
Feiner, Ralph 213
Frankl, Leo 56
Giorgi, Stefania 208, 209
Greb, Anneliese 142, 143
Henrikson, Gunnar 146, 147
Hoppler, Käthi + Ruedi 167, 168,
169
José, Jean François 20, 21
Kane, Joanna 67, 68, 69, 70, 71
Klenzendorf, Christoph 229
Knuchel, Hans 121
Küenzi, Adrian 90, 91
Küenzi, Sasha 87, 89
Kunert, Frank 149
Kunkovàcs, Làszló 185, 187, 188
Lindner, Ute 117
Mertens, Kerstin 135, 136, 137
Paakkunainen, Ulla 193
Quentin, Robert 177, 178, 179
Staatliche Eremitage, St. Peters-
burg 54
Steets, David 57
Thomas, Katharina 54, 55, 246
van S, Mark 248
© VG Bild-Kunst Bonn 2000 38,
39, 146, 147, 163, 165, 166**

Namens-Register
Alle Ziffern beziehen sich auf die jeweilige Seite des deutschen Textes

Index of Names
All numbers refer to the respective pages of the German texts

© 2000 ARNOLDSCHE und Autoren

Autoren Authors
Peter Schmitt M. A.
Dr. Kirsten Claudia Voigt
Sabina Heck
Katharina Thomas
Inge Evers

Redaktion Editorial work
Peter Schmitt M. A.
Katharina Thomas

Katalogbearbeiter
Editor of Catalogue
Peter Schmitt M. A.

Graphik Design
Silke Nalbach, Stuttgart

Offset-Reproduktionen
Offset Reproductions
Konzept Satz und Repro, Stuttgart

Druck Printing
Rung-Druck, Göppingen

Dieses Buch wurde gedruckt auf 100 % chlorfrei gebleichtem Papier und entspricht damit dem TCF-Standard
This book has been printed on paper that is 100 % free of chlorine bleach in conformity with TCF standards

Die Deutsche Bibliothek – CIP-Einheitsaufnahme

Ein Titeldatensatz für diese Publikation ist bei der Deutschen Bibliothek erhältlich

ISBN 3-89790-157-9

Made in Europe, 2000